KRISTINA DIPROSE
GILL VALENTINE
ROBERT M. VANDERBECK
CHEN LIU
AND KATIE MCQUAID

CLIMATE CHANGE, CONSUMPTION AND INTERGENERATIONAL JUSTICE

Lived Experiences in China, Uganda and the UK

BRISTOL
UNIVERSITY
PRESS

First published in Great Britain in 2019 by

Bristol University Press
University of Bristol
1-9 Old Park Hill
Bristol
BS2 8BB
UK
t: +44 (0)117 954 5940
www.bristoluniversitypress.co.uk

North America office:
Bristol University Press
c/o The University of Chicago Press
1427 East 60th Street
Chicago, IL 60637, USA
t: +1 773 702 7700
f: +1 773 702 9756
sales@press.uchicago.edu
www.press.uchicago.edu

British Library Cataloguing in Publication Data
A catalogue record for this book is available from the British Library.

Library of Congress Cataloging-in-Publication Data
A catalog record for this book has been requested.

ISBN 978-1-5292-0473-5 (hardback)
ISBN 978-1-5292-0475-9 (ePub)
ISBN 978-1-5292-0474-2 (ePDF)

Cover design by blu inc, Bristol
Front cover: image kindly supplied by Katie McQuaid
Printed and bound in Great Britain by CPI Group (UK) Ltd, Croydon, CR0 4YY
Bristol University Press uses environmentally responsible print partners

Contents

List of Figures and Photographs

Figures

Photographs

Notes on the Authors

Kristina Diprose is a research associate in the University of Sheffield's Urban Institute. She is interested in social research, knowledge coproduction and creative practice towards the creation of 'just' cities, with a particular focus on climate change and sustainability.

Chen Liu is a lecturer in Cultural Geography at the School of Geography and Planning, Sun Yat-sen University. Her research interests include food consumption, popular culture and everyday life in urban China. She has published more than 15 articles and book chapters in both English and Chinese.

Katie McQuaid is a senior research fellow in the School of Geography at the University of Leeds. Her research interests include gender and sexuality, local perspectives on climate change, (inter)generational relations in informal urban African settings, and the intersection of applied arts and feminist methodologies in action research with marginalized communities.

Gill Valentine is Professor of Human Geography at the University of Sheffield. Her key areas of research include equality, diversity and inclusion; childhood, parenting and family life; and urban cultures and consumption. She is a fellow of the Academy of Social Sciences and has published 15 books and over 150 articles.

Robert M. Vanderbeck is Professor of Human Geography at the University of Leeds, with particular expertise in childhood, youth and intergenerational relationships; sexuality and religion in the public sphere; and contemporary processes of social inclusion and exclusion. He is the author of *Law, religion* *and homosexuality* (with Paul Johnson, Routledge, 2014) and co-editor of *Intergenerational space* (with Nancy Worth, Routledge, 2014).

Acknowledgements

This book draws on some material from previously published articles, but crucially brings research data from three cities together for the first time with original synthesis of research findings. We acknowledge the following team publications that have contributed to the development of this book, with material reprinted with permission from: 'Building common cause towards sustainable consumption: A cross-generational perspective' (Diprose et al, 2019a); 'Caring for the future: Climate change and intergenerational responsibility in China and the UK' (Diprose et al, 2019b); 'Placing "sustainability" in context: Narratives of sustainable consumption in Nanjing, China' (Liu et al, 2018a); 'Intergenerational community-based research and creative practice: Promoting environmental sustainability in Jinja, Uganda' (McQuaid et al, 2017); and 'Urban climate change, livelihood vulnerability and narratives of generational responsibility in Jinja, Uganda' (McQuaid et al, 2018).

We would like to thank the following people for their contributions to the INTERSECTION research programme: co-investigators Dr Lily Chen and Dr Mei Zhang from the School of East Asian Studies at the University of Sheffield; intergenerational theatre lead Professor Jane Plastow at the Leeds University Centre for African Studies; theatre co-facilitators Baron Oron in Jinja, Ping Chen, Ziyi Li and Kai Yu in Nanjing, and Matthew Elliot in Sheffield; local collaborators We Are Walukuba, Passages Theatre Group, Sheffield College Performing Arts, and Nanjing Arts Institute; and Gravel & Sugar filmmakers Valentina Tschismarov, Matylda Wierietielny and Will Nyerere. Thanks also to Sheffield sculptor Anthony Bennett for his work on the *Sustainability Dancer*, to Helen Mort for facilitating the *Write About Time* workshop, to former INTERSECTION research staff Catherine

Harris and Lucy Jackson, and to the INTERSECTION Advisory Board: Ade Sofola (formerly Children England), Antony Mason (Intergenerational Foundation), Elaine Willis (formerly Beth Johnson Foundation), Fiona Matthews (Earth Champions Foundation), Frances Babbage, Maria Grasso and Matt Watson (University of Sheffield), Luisa Golob (Ignite Imaginations), Peter Hopkins (Newcastle University) and Sue Mayo (Magic Me).

Select photos from INTERSECTION discussion groups, intergenerational theatre workshop events and of the public artwork *Sustainability Dancer* are included in the photo spread between Chapters Three and Four.

ONE

Introduction

As we finish this book, schoolchildren are organizing mass walkouts under the banner of 'Youth Strike 4 Climate' as part of a global campaign for action on climate change. The movement, which began with 15-year-old Greta Thunberg protesting outside the Swedish Parliament, is the latest to draw attention to the intergenerational injustice of climate change, with today's politicians and adults blamed for short-changing the future. 'We are going to have to pay for the older generation's mistakes', one of the school strikers told the BBC (2019). In diverse international contexts, there is growing concern that prior and current generations have made choices that will severely curtail the ability of future generations to pursue their interests and to lead liveable lives (Vanderbeck and Worth, 2014).

Whether knowingly or not, the arguments put forward by these young protesters echo arguments for sustainable development that first rose to prominence in the late 1980s, such as those advanced by the publication of the Bruntland Commission report *Our Common Future*. Bruntland's definition of sustainable development as 'development that meets the needs of the present without compromising the ability of future generations to meet their own needs' (WCED, 1987: 16) has become a common touchstone for sustainability advocates (Agyeman et al, 2002). Developing in parallel to sustainable development discourses, public debate on climate change has often focused on the need to safeguard the future, for example in the stark warning from Christine Lagarde, head of the International Monetary Fund, that 'unless we take action on climate change, future generations will be roasted, toasted, fried and grilled' (citied in Marshall, 2014: 29). These arguments are familiar, but coming from young people they sound unrehearsed, angry, and more urgent.

As Greta's one-girl school strike became international news, inspiring a movement of youth-led protests from Australia to Uganda, a new report from the United Nations' Intergovernmental Panel on Climate Change (IPCC) warned that the future catastrophe these young people hope to avoid is closer than we may imagine. The IPCC estimates that the global temperature has already risen by 1°C above pre-industrial levels as a result of human activity (IPCC, 2018). Its experts warn that global warming is likely to reach 1.5°C by as early as 2030, well within the lifespans of most adults alive today as well as those of the schoolchildren protesting our inaction.

Writing on climate change in 1989, Bill McKibben warned that it is a mistake to believe that nature 'moves with infinite slowness', proclaiming that 'our reassuring sense of a timeless future ... is a delusion' (McKibben, 2003 [1989]: 5). Since the advent of industrialization – a relatively recent phenomenon, ecologically-speaking – and the emergence of mass-production and consumption as the *modus operandi* for development, human activity has caused global environmental shifts of 'an unprecedented scale and speed' (Hamilton et al, 2015: 4). Climate change is an intergenerational moral challenge that concerns our ability to make decisions in the interests of the future, but it is also happening now. The IPCC's report records evidence of human-induced changes in the frequency and intensity of climate and weather extremes since the 1950s. It has been evident for decades that there is a link between climate change and phenomena such as hurricanes, droughts, wildfires, flooding and sea level rise, with consequent implications for biodiversity and ecosystems, species loss, and for human health, livelihoods, food security, water supply, and safety. Yet climate change is commonly cast as 'distant and abstract' (Klein, 2015: 3), as something to worry about later.

The ability to distance ourselves from climate change in this way is not universal. Its effects are felt to a greater or lesser degree depending on where in the world we live, and the extent to which our livelihoods directly rely on favourable climatic conditions. The fact that climate change is mostly caused by the activities of rich, industrialized nations and mostly impacts poor, less industrialized nations links environmental concerns to questions of social justice. Calls for climate justice typically focus both on intergenerational and *intra*generational equity between the major polluters and victims of climate change (Agyeman et al, 2002). Ugandan President Yoweri Museveni has, for example, described climate change as 'an act of aggression by the rich against the poor' (cited in Jamieson, 2010: 438). Climate change is a prime example of the difficulty we face in acting ethically when the scale of human influence is extended all over the world and far into the future (Persson and Savulescu, 2012), especially for those who lead a relatively 'charmed' existence while 'the people and the pollution that sustain us are invisible to us' (Pearce, 2008: 4).

A large academic literature explores the impact of climate change and the arguments we have just outlined for intergenerational and intragenerational justice. Much of this literature is written from the perspective of what climate science can tell us, or else it outlines legal, philosophical and political approaches for addressing climate change, or offers case studies of how communities on the frontlines of climate change are adapting. This book is, like the latter, grounded in empirical social research. It is not about adaptation per se, but rather how ordinary people in different and power-differentiated parts of the world are responding to 'deep questions about how we should live on the planet' (Castree, 2017: 55). In particular, it is about the extent to which people's lived experiences of climate and social change reflect and complicate considerations of intergenerational and intragenerational justice. This work is situated within a recent turn towards understanding the 'human sense of climate' (Hulme, 2017), drawing on anthropological insights about the relevance of local environmental knowledge for interpreting climate change and motivating action.

About INTERSECTION

This book is based on a three-year interdisciplinary research programme called INTERSECTION, which was funded by the Arts and Humanities Research Council's Care for the Future programme from 2014 to 2017. INTERSECTION was an innovative, cross-national programme that employed participatory arts and social science methods to explore the themes of intergenerational justice, consumption and sustainability with urban residents in three cities: Jinja in Uganda, Nanjing in China, and Sheffield in the United Kingdom. These cities are very differently positioned in relation to global networks of production and consumption, processes of (de)industrialization, and vulnerability to climate change; and also in relation to the composition and spacing of generations, and cultural understandings of familial and intergenerational obligations.

Our intention in working across cities in China, Uganda and the UK was to bring Global North perspectives on responsibility for climate change and sustainable consumption into dialogue with those of citizens in more recently industrialized and emerging economies. The UK is a significant player in the history of fossil fuel extraction through early industrialization, while China is the foremost contributor to present day global greenhouse gas emissions (Cuomo, 2011; Yu, 2014; Hansen and Sato, 2016). Uganda is ranked among the world's least developed countries (UN/DESA, 2018) and also among those most vulnerable to the impact of climate change (ND-GAIN, 2019). The majority of the UK and China's population live in urban areas, while Uganda's population is predominantly rural but rapidly urbanizing. The juxtaposition of the former 'Steel City' of Sheffield with the fast-growing cities of Jinja and Nanjing reflects shifting global concentrations of mass-production, consumption and associated environmental problems. In the next section we briefly outline some notable features of each city.

Case studies

Uganda, in sub-Saharan Africa, is among those nations especially vulnerable to climate change. Even in urban areas such as Jinja – an industrial city of 80,000 people on the northern shore of Lake Victoria – many residents rely directly on agriculture or peri-urban agriculture for subsistence, have direct links to family members living in rural areas, and buy food from local farms. During our fieldwork there in both 2015 and 2016 drought was affecting many parts of the country, causing crop failure, and driving up food prices and other living costs. High consumption lifestyles have only recently started to emerge, reflecting economic growth, urbanization and the increasing influence of global media. (Inter)generational considerations are critical as the country has the world's second-youngest age structure. Many older and newer urban residents from subsistence backgrounds are not used to cash economies, and this is reflected in anxieties about intergenerational value change and rising materialism in the younger 'dot com' generation. Intergenerational relationships and imaginaries of the future have been significantly affected by high HIV/AIDS mortality rates, as well as by rising informality and unemployment among the urban poor.

China occupies a pivotal role in debates about responsibility for climate change, as the world's largest greenhouse gas emitter following its rapid industrialization and growth in the post-Reform era. The family planning policy introduced in the late 1970s has generated changes in kinship structures and contributed to the 'Little Emperors' phenomenon: a young, emerging urban middle class who enjoy a higher standard of living than previous generations and place increasing emphasis on consumption. Sustainable development is becoming a national priority, reflected in the official Chinese discourse of *shēngtài wénmíng* (生态文明, 'ecological civilization'). Environmental problems such as air quality are a major concern for many urban residents. Our fieldwork in China focused on Nanjing, a major industrial and commercial city of over 8 million people and the capital of Jiangsu province in the Yangtze River Delta region. Nanjing was one of the worst affected cities in the 2013 Eastern China Smog, which caused significant disruption to transport, school closures, and a spike in respiratory illnesses.

Sheffield in South Yorkshire, UK, is home to more than half a million people within a wider metropolitan district of 1.5 million. It typifies the transition of many Western European cities from a landscape of production to consumption of cultural, service, retail and digital industries. With the post-war decline in manufacturing and intensification of consumerist economics has come concern about whether or not this model is sustainable and its impact on quality of life. This was exacerbated by the 2008 economic crisis and subsequent recession and national austerity programme, which brought focus to a possible intergenerational decline in living standards between 'Baby Boomers' and 'Millennials'. The UK's history of early industrialization and resource extraction make it a major contributor to global climate change. Nationally, there has been some recent progress in reducing domestic greenhouse gas emissions, but the country is nonetheless exposed to the global trend towards more extreme weather. Extensively urbanized and situated at the confluence of five rivers, Sheffield is particularly flood vulnerable and last experienced major flooding in 2007.

Fieldwork and focus of this book

INTERSECTION fieldwork was undertaken between January 2015 and December 2016 by an embedded researcher in each of the three participating cities. This encompassed quantitative and qualitative social research and creative practice focused on the key programme themes of intergenerational justice, consumption, and sustainability. We were interested in local experiences of climate change, as well as perceptions of intra- and intergenerational equity in current consumption trends, seeking to understand how people perceive:

- Who is entitled to what?
- Who wins and who loses?
- Who owes what to whom?

Through these questions, we explored cultural and generational perspectives on entitlement to consumption, who is to blame for unsustainable consumption, and who ought to take responsibility for the environment that future generations will inherit. Mindful of the way that climate change is often cast as a generational battleground, we were also eager to provide opportunities through our research for people of different generations to come together.

The social research approaches that have informed the writing of this book include:

- a cross-generational survey of 750 urban residents in each city;
- narrative interviews with a cross-generational sample of urban residents (circa 90 individual interviews in each city);
- three-generation family interviews, typically inclusive of young adults, a parent and grandparent, and sometimes extended family (circa 15 families in each city); and
- generational dialogue groups, which brought together broad generational cohorts of younger, middle-aged and older people (circa nine groups in each city).

Additionally, we organized intergenerational community theatre workshops in each city and, through collaborations with filmmakers Gravel & Sugar, sculptor Anthony Bennett and poet Helen Mort, also organized a series of creative public engagement events. This book is primarily based on social research data from the survey, 400+ interview transcripts and generational dialogue groups, with some case study material on creative practice in the final chapter. It draws on the co-authors' expertise in social, cultural and intergenerational geographies and in anthropology, though it is intended for a wide readership across related environmental social research disciplines, and we hope will also be of use to policymakers and practitioners interested in public perceptions of climate change, creative public engagement and intergenerational practice.

More information about INTERSECTION, including links to our other publications, creative work, short films and the Gravel & Sugar documentary *Osbomb, love and Supershop: Performing sustainable worlds* (2017), can be found on the programme website: www.sheffield.ac.uk/intersection.

Note on cross-cultural research

Johnson (1998: 1) observes that '[i]n perhaps no other subfield of social science research are issues of methodology and measurement as open to challenge and criticism as when they are applied in cross-cultural and cross-national settings.' The INTERSECTION programme's focus on cities spanning three countries and continents necessitated serious consideration of issues of cross-cultural comparability in the design, collection and analysis of data. Given our interest in everyday perceptions of large and abstract concepts like climate change, (un)sustainable consumption and intergenerational justice, our biggest challenge was the issue of 'conceptual equivalence' (Bulmer, 1998; Johnson, 1998; Johnson and van den Vijver, 2003; He and van de Vijver, 2012). This meant considering the language and framing employed in our interview guides and survey, reflecting critically at each stage of the research process on 'whether the terms one is using … mean the same in different societies? … Do literally equivalent words and phrases convey equivalent meanings?' (Bulmer, 1998: 160–61). This is not only an issue of translation, but also knowledge of local cultural contexts.

Each of our research instruments was co-designed by a research team that included UK and Chinese scholars and a dedicated researcher embedded in each city. The Jinja Research Associate, Dr Katie McQuaid, had significant prior experience of ethnographic fieldwork in Uganda. The mixed-methods research design and cultural immersion of research team members within each local urban context allowed us to develop 'interpretively equivalent' lines of questioning (Johnson, 1998); while the simultaneous development and discussion of research instruments across different sites enabled a degree of 'cultural decentering' (Werner and Campbell, 1970; He and van de Vijver, 2012). Where appropriate, we adapted and piloted questions to find cross-cultural common ground. Where differences in cultural interpretation of key terms were integral to our research, we noted and reflected on this in our analysis. In Chapter Two, we discuss how the cultural framing of climate change (*qìhòu biànhuà* or 气候变化 in Chinese) differed across the three cities, and the importance of contextualising survey findings on the extent to which urban residents are concerned about climate change alongside qualitative data on these differences in perception.

Initial drafts of the survey and interview guides in English included questions about sustainable consumption. Local researcher feedback and pilot research from both Jinja and Nanjing revealed this to be an unhelpful term to use. In Jinja it meant little to local residents, reflecting its limited applicability as a concept originating in response to the problem of extravagant consumption within particular societies. In Nanjing, its direct translation *kě chíxù xiāofèi* (可持续消费) was perceived as the 'official' language of the state, rather than a term that people would use in relation to their everyday consumption practices

(Thøgersen, 2006). While we retained an interest in this topic, we found alternative ways to ask urban residents about 'things you buy and use' and their environmental impact (or, in Chinese, 'sustainable ways to buy and use things', *kě chíxù gòumǎi hé shǐyòng wùpǐn de fāngshì*, 可持续购买和使用物品的方式). Similarly, the terms sustainability and intergenerational justice were perceived as technical and abstract, proving off-putting to residents across all three cities in pilot research. Rather than employ these terms directly, we used related concepts such as what people consider to be a 'fair' (*gōngpíng* or 公平) use of resources between nations and generations, what impact they think climate change has now and will have in the future, who they think is to blame for this, who they think is responsible for conserving the environment, and what they think those responsible should do.

We also had to consider the possibility of cultural differences in response styles in the context of power-differentiated research encounters (Johnson, 1998; Johnson and van de Vijver, 2003; He and van de Vijver, 2012). This can include varying degrees of acquiescence (the tendency to agree rather than disagree with an interviewer) and social desirability bias, whereby people self-censor to conform with perceived social norms. Although these are universal tendencies, research suggests that they are present to a greater or lesser degree across different cultures (Johnson and van de Vijver, 2003). Social desirability bias has been found to be more prominent in collectivist countries, where concerns about maintaining good relationships and saving face are more salient (Triandis, 1995). Researchers have found that this intersects with a 'courtesy bias' in traditional Asian cultures, for example suggesting that Chinese respondents may pretend to be environmentally concerned even when they are not (Jones, 1963; Chan and Lau, 2000; Keillor et al, 2001; Johnson and van de Vijver, 2003). Acquiescence is more common among people with low socioeconomic status from collectivist cultures (Harzing, 2006; Smith and Fischer, 2008; He and van de Vijver, 2012); while 'people coming from more influential groups in society or from more affluent countries' tend to be less concerned with creating a favourable impression (Johnson and van de Vijver, 2003). Moreover, Bulmer (1998) highlights different levels of familiarity with exported social research methods that put Western respondents at an advantage, as they live in societies that are 'saturated' with survey research and interviews, and tend to be more familiar with how the data is likely to be used and interpreted. At various points in this book, we draw on survey data where cross-cultural comparison may to some extent be distorted by these issues. However, this data is always contextualised in relation to in-depth qualitative social research and ethnography, rather than presented as standalone evidence of cultural difference (Bulmer, 1998; Johnson, 1998).

Nanjing interviews and focus groups were conducted and transcribed in Mandarin Chinese and then translated into English. As with any social research in translation, there may be some loss of 'original' meaning, associations and contexts, inevitably sacrificing some of the 'situatedness' and specificity of the Nanjing data when presented in English (Smith, 1996; Desbiens and Ruddick, 2006; Müller, 2007). Whilst the act of translation means accepting 'incompleteness' and acknowledging these limitations (Bialasiewicz and Minca, 2005), movement between languages can also give rise to 'in-between' forms of understanding and openness to cross-cultural communication (Smith, 1996).

While a number of respondents in the centre of Jinja, and many younger people generally, spoke English, some older people, rural to urban migrants and those living in informal settlements did not. In these cases, we engaged the services of an experienced translator from Makerere University who grew up in Jinja. Luganda and Lusoga were the two most commonly spoken languages where English was not known. In exceptional circumstances, when a member of the older generation spoke a language unknown to the translator, a member of their family would provide assistance. Borchgrevink (2003) provides an instructive account of the challenges of using interpretation in ethnographic work. In recognition of deeply rooted patriarchal ideologies and gender dynamics in this region of Uganda (Ochieng, 2003; Tamale, 2003; Hayhurst, 2014), a female translator was chosen.

Ethical practice was of paramount importance across all three cities, particularly in light of cross-cultural differences in interview dynamics and settings. Our work adhered to University and Research Council ethical principles, as well as those of key local gatekeeping institutions including, for example, the Uganda National Council for Science and Technology. As part of our commitments, names in this book have been changed to protect the anonymity of research participants.

TWO

A Global and Intergenerational Storm

Introduction

This chapter seeks to situate the INTERSECTION programme of research within wider international debates regarding the relationship between consumption and climate change. We consider how this relationship is addressed in arguments for environmental justice and sustainable development, and how it is reflected in international policymaking. Our discussion highlights how climate change is typically cast as both an international and intergenerational injustice, or the convergence of a 'global storm' and an 'intergenerational storm' (Gardiner, 2006). This demonstrates the need for research with both a cross-national and cross-generational perspective that explores people's lived experiences of climate change and how they relate this to their own and other people's historical and contemporary patterns of consumption. We situate our research within recent social science scholarship that explores how people live with a changing climate, advocating a 'human sense' of climate and social change (Hulme, 2017). Finally, we introduce the main themes and original contribution of our work in Jinja, Nanjing and Sheffield with an overview of the subsequent chapters, which are based on the analysis of a unique body of primary data.

Consumption and climate change

Since the advent of the industrial revolution, unsustainable forms of consumption have contributed to global environmental degradation and climate change. The link between consumption and climate change is well-established, with overwhelming evidence of unprecedented levels of anthropogenic greenhouse gas emissions from fossil fuels largely attributed to global economic growth, driven by mass-production and consumption (IPCC, 2014: 4). Natural and social scientists have argued that the scale of human impact on the planet is now so significant that it constitutes a new geological epoch, the Anthropocene, which represents 'a threshold marking a sharp change in the relationship of humans to the natural world' (Hamilton et al, 2015: 3). To illustrate just how transformational industrialization has been of our capacity to consume natural resources, Berners-Lee and Clark (2013) observe that annual energy consumption today is equivalent to each person on the planet having more than 100 full-time servants doing manual work on their behalf.

Before the impact of climate change was fully understood, the assumption was that ever-escalating growth in production and consumption was both necessary and desirable for development. As national governments pursue growth, carbon-intensive consumer lifestyles become possible for more people worldwide (Meyers and Kent, 2003). Yet the overwhelming evidence from advanced industrialized nations is that these lifestyles are unsustainable. This is well illustrated by the widely circulated statistic that if everyone had the ecological footprint of the average United States citizen, we would need four planets to sustain us instead of one (McDonald, 2015). This highlights how some people take up far more than their fair share of 'environmental space', enjoying high consumption lifestyles while extracting labour and resources from, and displacing pollution to, poorer parts of the world (Pearce, 2008; Agyeman, 2013). This raises profound questions of intra- and intergenerational justice in relation to current consumption trends in long-established high income countries and consumption aspirations elsewhere. Jackson (2009) suggests that for living standards worldwide to be equivalent to those in Western Europe by 2050, we would need to increase our technological efficiency by an unprecedented 130-fold. Nair (2011) similarly argues that if Asia were to achieve consumption levels taken for granted in the West, the results would be environmentally catastrophic. There is also emerging evidence from high income, high consumption countries that beyond a certain threshold, the relationship between economic growth and human wellbeing is weak at best. Therefore, it is argued, transitioning to less competitive and more equitable forms of consumption might offer a 'double dividend', proving better in the long run for people and the planet (Jackson, 2005, 2009; Pickett and Wilkinson, 2009; Agyeman, 2013).

The global picture

Until very recently, greenhouse gas emissions were accelerating globally in line with per capita GDP, with the growth rate strongest in rapidly developing economies such as China (Raupach et al, 2007). In the past decade, growth in carbon dioxide (CO_2) emissions – the biggest contributor to human-induced climate change – had slowed, and then stagnated in the years 2014–16. This was attributed to a switch to less carbon-intensive forms of energy, particularly the decreasing use of coal (Olivier et al, 2017), leading to hopes of stabilising global emissions. However, the most recent data offers less cause for optimism. Most countries' economies are still predominantly reliant on fossil fuels and, after a three-year hiatus, CO_2 emissions rose by 1.6 per cent in 2017, with a projected rise of 2.7 per cent in 2018 and (at the time of writing) further increases projected for 2019, driven by both industrialized and emerging economies (Le Quéré et al, 2017; Jackson et al, 2018). Technological innovation in recent decades appears to have had little impact in slowing or reversing the overall trend (Berners-Lee and Clark, 2013).

Even if global emissions were to stabilise, current levels are far from sustainable, and much of the damage is already done and will persist for many centuries (Hamilton, 2010; McKibben, 2012). Based on current trends and likely future scenarios, the IPCC forecasts that the global mean surface temperature could rise by 1.5°C above pre-industrial levels by as early as 2030, and by more than 2°C by the end of the 21st century. 1.5°C is increasingly recognized by climate scientists as a 'safe' global limit, beyond which the impacts of global warming 'may be long lasting and irreversible' (IPCC, 2014, 2018). This unnatural level of warming disrupts ecosystems at a challenging pace, threatening countless species and human livelihoods, with major implications for the future.

Environmental justice and sustainability

Agyeman et al (2002) trace the emergence of two distinct approaches to the problems caused by inequitable resource consumption and its present and future impact on the environment. The first of these, sustainability, rose to prominence as an international policy agenda with the publication of the 1987 Bruntland Commission report *Our Common Future* and the landmark 1992 Rio Earth Summit, which brought to a head decades of rising environmental concern around the world (Hamilton, 2010; Agyeman, 2013). In response, the Summit established the United Nations Framework Convention on Climate Change (UNFCCC) and released Agenda 21, which stated that national governments should maintain 'sustainable livelihoods for the south' and 'sustainable production and consumption for the north' (Hobson, 2002: 98). More recently, the UN's 2030 Agenda for Sustainable Development, in particular Sustainable Development Goals 12 and 13, focus on sustainable production and consumption and climate change as pressing, interlinked challenges for both the Global North and South (UN General Assembly, 2015). Agyeman et al characterize this sustainability agenda as 'top-down' and 'futures-orientated', having emerged 'in large part from international processes and committees, governmental structures, think tanks and international NGO networks' (2002: 88) as a response to the problem of living within planetary limits.

Environmental justice, on the other hand, originated in grassroots civil rights activism in the United States as an organized response to environmental racism, in protest against the disproportionate impact of pollution and toxic waste on poor, largely African-American communities (Bullard, 1990; Agyeman et al, 2002; Walker, 2011). As such, it places greater emphasis on the relationship between environmental and social inequality, focusing on the distribution of environmental 'goods' and 'bads' and inequitable access to decision-making processes. As the movement and concept has travelled beyond the US, it has expanded to encompass other forms of social difference such as class, gender, indigeneity and disability, and it has moved across scales to include global concerns like climate change (Walker, 2011). Environmental justice movements have been influential in reframing what were previously seen as 'green' agendas within a broader discourse of justice, rights and equity (Agyeman, 2000). In response, Agyeman (2013) argues, sustainable development discourses are beginning to reflect a more holistic set of 'just sustainability' concerns: improving quality of life, meeting the needs of present and future generations, access to justice and inclusion in decision-making processes, and living within ecosystem limits.

International climate change policy offers a prime example of attempts to incorporate considerations of environmental justice at a global scale (Agyeman, 2013), with the UNFCCC formally recognizing 'common but differentiated responsibilities' for climate change (UNFCCC, 1992: 1; Caney, 2005). This principle holds that industrialized countries have polluted more and are better resourced to address climate change, and should therefore be expected to shoulder more of the burden of emission control, and should also compensate developing countries through investment in their adaptive capacity (Den

Elzen and Schaeffer, 2002; Müller et al, 2009). Similarly, the 'polluter pays principle' holds that those who produce pollution should bear the cost of managing it to prevent damage to human health and the environment (Ward and Hicks, 2014). These principles underpin international policy mechanisms such as the Green Climate Fund, through which wealthier, advanced industrialized countries within the UNFCCC pledge funds to support climate-resilient development in those countries that are most vulnerable to climate change, in particular least developed countries, low-lying small island states and African states (Green Climate Fund, 2018). UN Sustainable Development Goal 13 targets call on developed countries to jointly mobilize $100 billion annually through the Green Climate Fund by 2020, and to promote mechanisms for raising capacity for effective climate change-related planning and management in least developed and developing countries (UN General Assembly, 2015). These are important steps towards transitioning to a more sustainable and more equitable sharing of environmental space, however – based on current global greenhouse gas emissions trends – they are not enough.

An international and intergenerational injustice

Climate change has been described as an 'existential crisis for the human species' (Klein, 2015: 15), presenting 'an ethical issue of epic proportions, for it endangers everything on Earth that human beings depend upon and care about' (Cuomo, 2011: 692). It is commonly cast as an injustice on two fronts:

- the **international**: with less industrialized nations typically being both less culpable for and more vulnerable to the effects of climate change (Stern, 2007; Füssel, 2010; Jamieson, 2010; Hansen and Sato, 2016); and
- the **intergenerational**: with a time lag of centuries between past and present fossil fuel emissions and future ecological

harm (Gardiner, 2001; Caney, 2005; Hamilton, 2010; Baatz, 2013).

Gardiner (2006) argues that responses to climate change tend to address one or the other of these issues, which he refers to as 'the global storm' and 'the intergenerational storm'. This is well illustrated by recent emissions reduction pledges made through the UNFCCC's 2015 Paris Agreement. As outlined earlier, the focus of UNFCCC negotiations is the respective responsibilities of more and less industrialized nations for addressing climate change, with negotiators in theory pledging emissions reductions relative to the size of national economies and their environmental impact (Page, 2008). Analysis by the UN Environment Programme (UNEP, 2017) suggests that the pledges made in the Paris Agreement collectively account for only a third of the emissions reductions needed to avoid 2°C of global warming, an already lenient threshold. Moreover, these are voluntary rather than legally-binding targets. This shows how reaching agreement on a 'fair' global distribution of climate change mitigation actions can still spell disaster for future generations. As in previous UNFCCC negotiations, it appears that 'political realism bested scientific data' (McKibben, 2012).

To think through key features of the global and intergenerational storm, next we outline the relative positions of China, Uganda and the UK in terms of their vulnerability to and responsibility for climate change, and the ways in which intergenerational justice considerations feature in national public discourse.

The global storm

'The global storm' is the idea that the causes and effects of climate change are widely dispersed, that it is difficult to establish direct blameworthiness because the causes are collective and the harms diffuse, and that as a result of this it is difficult to coordinate an effective response (Gardiner, 2006; Jamieson, 2010). In spite of this, a particular moral framing of climate change has emerged in recent decades, drawing on research evidence that the countries that have contributed least to global greenhouse gas emissions are also the ones most vulnerable to the impacts of climate change (Samson et al, 2011; Hansen and Sato, 2016). This 'double inequity' (Füssel, 2010) is due to a range of factors, including: the geographic distribution and magnitude of climate change impacts; the extent to which people's livelihoods directly depend on favourable climatic conditions; the socioeconomic exposure of poorer nations to food security and human health challenges; and differences in adaptive capacity. Rudiak-Gould (2014: 366) identifies 'industrial blame' as an influential moral stance that frames climate change as 'a crime perpetuated by one human group upon another', usually characterized as Western, capitalist or industrial.

Regional risks

Different world regions have different levels of exposure to climate change risk (IPCC, 2014). In Europe many people see climate change as a distant and future threat (Marshall, 2014), but European settlements are exposed to damages from river and coastal flooding; water restrictions; and damage from heatwaves and wildfires. Flooding is a particular concern and priority in the UK (Committee on Climate Change, 2016). Asia is also vulnerable to flood damage and heatwaves, as well as drought-related water and food shortages. Rapid industrialization in China has sparked concerns over air and water quality and its impact on human health and quality of life (Holdaway, 2010; Ma, 2010; Li and Tilt, 2017). The African continent is most immediately vulnerable to climate change. Here it compounds existing stress on water resources, reduces crop productivity in communities that are directly dependent on agriculture for their livelihoods and food security, and encourages the spread of vector and water-borne diseases such as malaria. While much climate adaptation research has focused on rural communities, many urban residents in African cities and towns are similarly at risk (McQuaid et al, 2018a).

Complex geographies

Principles such as polluter pays appear to be a relatively straightforward way of addressing the global storm of climate change through an environmental justice lens, recognizing that industrialized nations owe an 'ecological debt' due to their overconsumption (McLaren, 2003). This is, however, complicated by the issue of historic responsibility for global greenhouse gas emissions and the interconnectedness of global trade, a challenge outlined by Raupach et al:

> Developed nations have used two centuries of fossil fuel emissions to achieve their present economic status, whereas developing nations are currently experiencing intensive development with a high energy requirement, much of the demand being met by fossil fuels. A significant factor is the physical movement of energy-intensive activities from developed to developing countries with increasing globalization of the economy. (2007: 10292)

Today China is the world's largest greenhouse gas emitter by a significant margin, accounting for a quarter of all global emissions, while the UK's contribution is estimated at 1 per cent and the whole of the African continent's share at 4 per cent (Hansen and Sato, 2016). The scale of 'the China problem' looms large over discussions of global climate change and is often invoked as rendering other nations' domestic emissions reduction efforts insignificant (Vandenbergh, 2008: 905). Yet when population size is taken into account, China's per capita emissions are roughly in line with those of the UK and the rest of Europe (Hansen and Sato, 2016; Le Quéré et al, 2017), while Uganda's are among the lowest in the world (UNDP, 2015). When cumulative historical CO_2 emissions are taken into account – from the industrial revolution to now – the US is responsible for over a quarter of global emissions, more than twice China's share. European emissions similarly account for a greater share of the global total in historical context, with the UK and Germany's cumulative emissions alone equalling China's (Berners-Lee and Clark, 2013; Hansen and Sato, 2016; Le Quéré et al, 2017).

In line with similar climate change mitigation efforts across much of Western Europe, the UK has achieved around a 40 per cent reduction in CO_2 emissions from 1990 levels (BEIS, 2017), yet has a huge legacy carbon footprint for its size (Berners-Lee and Clark, 2013). When population size is taken into account, the UK's cumulative per capita emissions may exceed even the US (Hansen and Sato, 2016). Thus the UK and China are both major polluters, but differently positioned in terms of their past and present responsibility for climate change, which is also bound up with the historical forces of exploitation, colonialism, wealth acquisition and economic growth (Cuomo, 2011).

A further issue to consider is the extent to which emissions reductions in countries like the UK represent real savings, or simply the relocation of production to countries like China. Berners-Lee and Clark (2013) argue that emissions reductions in one world region are not enough to keep fossil fuels in the ground, as they tend to get burned elsewhere. Meanwhile, UK residents enjoy the benefits of cheap consumer goods made in China. Studies estimate that around a third of Chinese emissions can be attributed to the production of exports for consumer markets elsewhere, largely in the US and Europe (Weber et al, 2008; Müller et al, 2009; Oxfam, 2015). This very brief overview of the challenges of calculating national contributions to climate change illustrates that intuitions about justice between nations and generations meet with very complex geographies.

Who is responsible?

Debates about responsibility for climate change tend to focus on countries and their national governments as the relevant unit of analysis. This makes for a compelling narrative, as statistics suggest that the G20[1] group of countries (with the largest national economies) are responsible for 78 per cent of global greenhouse gas emissions (Olivier et al, 2017). However, another feature of the global storm is what Gardiner (2006: 399) calls the 'fragmentation of agency': the idea that responsibility for climate change is shared by actors across various scales, from multinational corporations to city planners and ordinary citizens. In other words, the nation-state is one level of social organization that is relevant to addressing climate change, but it is not the only relevant actor. Jamieson observes:

> climate change is largely caused by rich people, wherever they live, and is suffered by poor people, wherever they live. We can attribute primary responsibility for climate change to the 500 million people who emit half of the world's carbon, but not all of them live in the rich countries of the North. (Jamieson, 2010: 439)

In light of this, Oberheitmann and Sternfield (2011: 209) suggest moving beyond statist conceptions of responsibility for climate change, towards a more 'cosmopolitan' view of responsibility that focuses on consumers and polluters: 'a move away from debates about which *countries* are responsible for the problem towards debates about which *people* are responsible.'

---

[1] The G20 members are Argentina, Australia, Brazil, Canada, China, France, Germany, India, Indonesia, Italy, Japan, Republic of Korea, Mexico, Russia, Saudi Arabia, South Africa, Turkey, United Kingdom, United States and the European Union.

Recent research has drawn attention to the impact of affluent and new consumers. A 2015 Oxfam study estimated that the richest 10 per cent of the global population are responsible for almost half of all lifestyle-related consumption emissions, while the poorest 3.5 billion people on the planet account for only 10 per cent of emissions. The vast majority of affluent consumers live in OECD[2] countries but this is gradually changing, with well over 1 billion new consumers from developing and transition countries having achieved purchasing power parity with the US (Myers and Kent, 2003: 4963). China in particular has a burgeoning new consumer market. Yet, there are significant within-country differences in income and per capita emissions. In both China and the UK, the per capita carbon footprint of the richest 10 per cent of households is at least four times higher than the bottom 50 per cent (Oxfam, 2015). Despite efforts to make 'green' consumption appealing to more affluent consumers, research suggests that even when people identify with pro-environmental values, their income level is the strongest predictor of their carbon footprint (Moser and Kleinhückelkotten, 2018). This highlights the intersection of environmental and social inequality at different scales.

Population growth is another prominent and often contentious concern in the climate change debate (Berners-Lee and Clark, 2013; IPCC, 2014). One aspect of this is the impact of rapidly expanding new consumer markets in countries like China and their implications for carbon-intensive sectors such as the meat and car industries (Myers and Kent, 2003). The other is land use pressures increasing the global poor's vulnerability to climate change, reflected for example in concerns about food security, freshwater access and the spread of vector and water-borne diseases (McDonald et al, 2011; McQuaid et al, 2018a). However, given the significant difference in the carbon footprint of the global rich

[2] The Organisation for Economic Co-operation and Development is an intergovernmental economic organization with 35 member countries.

and poor by a magnitude of up to 1000, and the concentration of high population growth rates in nations with very low emissions per person, it makes sense to focus on consumption rather than population as the key driver of climate change (Satterthwaite, 2009; Berners-Lee and Clark, 2013; Oxfam, 2015). This issue illustrates the convergence of the global and the intergenerational 'storms' insofar as long-affluent countries have had a much greater impact on climate change, while in other countries people are only just beginning to enjoy equivalent lifestyles, and an overwhelming majority of low income, low emission households have yet to access a 'fair share' of global environmental space (Myers and Kent, 2003; Satterthwaite, 2009).

The intergenerational storm

Many of the effects of climate change take a long time to be fully realised, well beyond human lifespans (Gardiner, 2006; Hamilton, 2010). This intergenerational time lag means that the climate change impacts we experience today are the result of greenhouse gas emissions dating as far back as the industrial revolution, while the full cumulative effect of today's global emissions will have a greater impact on future generations (McKibben, 2012). This means that the 'fragmentation of agency' problem has a temporal as well as a spatial dimension. Hamilton (2010: 25) argues that '[t]he lag between emissions and their effects on climate and the irreversibility of those effects makes global warming a uniquely dangerous and intractable problem for humanity'. Past generations who have released CO_2 and other greenhouse gases into the atmosphere are now dead; and it seems 'grossly unfair' to ignore the problem and leave it for future generations to solve (Page, 2008: 556). This points to people alive today taking responsibility for climate change.

Some argue that while the polluter pays principle makes sense now we know the harm caused by burning fossil fuels, it is more difficult to justify why past generations who were ignorant of the long-term effects of CO_2 are blameworthy, or why their descendants should take responsibility for emissions accumulated over many generations (Caney, 2005; Baatz, 2013). Moreover, the fact that (some) people alive today enjoy the benefits of resource-intensive consumption while the worst of its detrimental impacts are deferred to the future, means that each generation has little incentive to take action on climate change (Gardiner, 2006). Jamieson sums up this problem and its intersection with the global storm of climate change as akin to:

> asking people who are now living very well, who under many scenarios have adequate resources for adaptation, to buy insurance that will mainly benefit poor people who will live in the future in some other country; and to do this primarily on the basis of predictions about the future based on climate models, expert reports and so on. (Jamieson, 2010: 434)

How societies make decisions about and build institutions equipped to deal with the long-term future threat of climate change is thus a key challenge of sustainable development. Hamilton et al (2015: 11) argue that in the Anthropocene era, 'there is a question about our capacity to make decisions regarding events that are beyond the human experience', and that this demands 'decision processes where the Earth's future can no longer be unknown'.

The long threat of climate change

A view of climate change as a problem for the future is often reinforced by politicians who describe it as a threat to future generations, and this is reflected in public attitudes to climate change in both the US and UK (Marshall, 2014). However, recent social science scholarship suggests a discrepancy in the way people tend to think about the future and the 'long threat' of climate change (Dickinson, 2009). The prevailing geoscientific narrative of its future impact models scenarios that will unfold over decades and centuries, while most people make decisions and enact intergenerational responsibilities within more immediate timescales (Hulme et al, 2009; Brace and Geoghegan, 2010; Persson and Savulescu, 2012). Climate change models with abstract cut-off points like 2050 or 2100 are beyond the scope of everyday intergenerational concerns. At the same time, they curtail possibilities for cultivating longitudinal thinking. Girvan (2014) contends that the way the imagined future of climate change is framed often 'precludes acting as if there is a more sustained future to care about' because engagement with 'deep time' – the longitudinal timescale at which climate change occurs – is discouraged. Hulme et al (2009: 198) similarly argue that 'time-scape analysis of climate change is overdue', suggesting that most people conceive of the future at most two decades from the present. Recent social research on climate adaptation illustrates that even in communities especially vulnerable to climate change, residents tend to focus on relatively short time horizons rather than long-term planning for the future (Fincher et al, 2014).

Synchronic and diachronic intergenerational equity

Climate change is perhaps the foremost example of an issue that concerns *diachronic* intergenerational equity: resource conserving efforts by people alive today on behalf of future generations (Attas, 2009: 207s). However, consistent with a view of time that privileges human lifespans and the difficulty that people have imagining the long-term future, debates about intergenerational justice and resource consumption more commonly focus on *synchronic* intergenerational equity between people of different generations alive today. This is reflected, for example, in prominent social policy debates around issues such as housing, education and access to employment. Moreover, anxieties about environmental and social change are often reflected in narratives of generational difference that characterize birth cohorts in particular ways (Mannheim, 1952 [1923]; Inglehart, 2008; Vanderbeck and Worth, 2014). In the context of consumption, a predominant generational narrative is one of frugality versus excess, with younger generations often negatively stereotyped as more consumer-driven and environmentally destructive than their forebears (Robins, 1999; Evans, 2011a; Carr et al, 2012; Keith et al, 2014; Yu, 2014; McQuaid et al, 2018a). In the next section, we briefly outline contemporary discourses of generational change, synchronic and diachronic intergenerational equity across each of our national case study sites.

Intergenerational concerns in China

Consumption patterns in urban China have dramatically changed following the pace of economic development since the launch of the 'open and reform policy' (the Reform) in 1978. Urban consumer culture has changed from a monotonous and state planned model into a globalized and marketized formation in recent years (Yu, 2014), and this is associated with increasing individualization in Chinese society (Yan, 2009, 2010). Since the emergence of the open market and the rise of non-state enterprises, Chinese consumers have tended to purchase more discretionary, luxury items, global brands and fashionable products to forge their social identities and desired lifestyles (Yu, 2014). Internationally, the consumption aspirations of China's expanding middle class and post-Reform generations raised to pursue 'The China Dream' (Goodman, 2014) provoke anxiety about Chinese consumers' present and future impact on global climate change.

Moreover, because of the shrinking number of children in each family due to the family planning policy introduced in the late 1970s, the previous parent-centred family pattern has dramatically transformed into a child-centred one. This is known as the 'Little Emperors' phenomenon (Stephen Parker et al, 2004;

Cameron et al, 2013) *or sì-èr-yī zōnghézhèng* (综合症, '4-2-1 syndrome'), which refers to four grandparents and two parents pampering their family's only child (Jing, 2000; Kong, 2010). On the one hand, China's Little Emperors are said to grow up 'spoiled' and consume much more than previous generations (Shao and Herbig, 1994; Wang, 2009). On the other, empirical studies have shown that the post-Reform generations have faced a greater income gap and social inequality in education and job markets than their parents, despite having better living conditions on the whole (see, for example, Wu and Treiman, 2007; Pun and Lu, 2010; Kan, 2013; Best, 2014; Hu et al, 2017; Schucher, 2017). Kan (2013) argues that the *èrdài* (二代, 'second-generation') phenomenon – which refers to people's personal success often relying on their family background – is connected to increasing inequality and social polarization among Chinese youth.

In China, the notion of generation is strongly related to an ethic of care and responsibility in both familial and public discourses. Chinese scholars argue that people alive today are obliged to maintain resources, wealth and environmental quality for future generations (Liao and Cheng, 2004; Zhang and Ruan, 2005; Fu, 2007). Contemporaries who occupy dominant natural, sociocultural, economic and political resources – typically working-age adults – are obliged to support elder generations and children (Jing, 2000; Kong, 2010). Fu (2007) argues that supporting the elderly based on a *huíbào* (回报, 'repay') or *huíkuì* (回馈, 'feedback') principle exemplifies Chinese understandings of fairness and equality. The traditional ideal of intergenerational interdependence appeals to parents and children to have a close relationship based on mutual support and reciprocity (Liu, 2017). Materially, housing arrangements often tie people from different generations together (Li and Shin, 2013). Emotionally, the responsibility of parents in cultivating and educating their children, and children's filial piety towards their parents, is considered important in Chinese families (Shek, 2006). Care for older and younger generations within the family exemplifies idealized intergenerational relations in the wider society. Confucius doctrine indicates that love and care begin within the family and irradiate outwards (Tuan, 1989; Fu, 2005). A widespread saying from Mencius (the successor of Confucius) states that a proper person should 'respect older people as we respect our own aged parents and care for the younger generation as we care for our own children' (Birdwhistell, 2007: 56).

Intergenerational concerns in Uganda

In Luganda (one of Uganda's languages, spoken by many of our research participants) the word for political regime and generation are the same: *omulembe*. Generations are named – if at all – after the political regime in power at the time: the current generation are named *omulembe gwa Museveni* ('generation/regime of Museveni'). Since Museveni assumed power in 1986, there has been relative stability (with the exception of civil war in the north) and advancements in women's rights, education and economic development. It is within this generation that current population growth has occurred, and within which the significant proportion of Ugandan youth have grown up. Many of the younger generation refer to themselves as the 'dot com' generation, which coincides with the *omulembe gwa Museveni* and the expansion of educational opportunities, particularly those who have accessed education through the Universal Primary Education (UPE) initiative. The 'dot com' generation is defined in large part by their education, but also the opening up of opportunities to connect with and learn about the wider world through access to modern technology and the internet. Those who describe themselves as part of this generation quip that older generations are 'BBC': born before computers.

In everyday life, people refer largely to only two generations: the older generation and the younger generation, with little concept of a middle-age bracket. Within these generations are four categories or 'social age cohorts' that span a socially and relationally constructed lifecourse: children and youth make up the younger generation and adults and elders make up the older generation. Research across sub-Saharan Africa has demonstrated the highly relational nature of age transitions, embedded in social relations with family members, peers, clans, ethnic groups and others in the community (Punch, 2002; Langevang, 2008; Evans, 2011) and evoking a 'social landscape of gendered power, rights, expectations and relationships' (Durham, 2000:116). Age transitions are understood as 'changes in the lives of individuals that are in accordance with the socially constructed life course' (Dewilde, 2003: 118), linked to normative 'milestones', such as leaving school, having children and marriage (Evans, 2012: 826). This highlights the diverse and often contradictory nature of young people's pathways to 'adulthood' in both the Global North and South (Valentine, 2003; Christiansen et al, 2006).

A child becomes a youth largely once they have finished schooling, but this is a slippery distinction as many never finish education, or do so at a wide range of ages. The main distinction between youth and adults is the taking up of responsibilities, principally through childbirth and marriage. An adult has responsibilities not just for their immediate family – their spouse, children and parents – but also their extended family and members of the clan (McQuaid et al, 2018b). However, in sub-Saharan Africa, difficult and constraining economic conditions often appear to foreclose the futures of young people who are portrayed as being 'stuck' (Sommers, 2011) or 'waiting' in youthhood, unable to gain employment, education, prestige or capital to transcend to adulthood (Honwana, 2012; Mains, 2012). The perceived inability of large numbers of African youth to achieve the social status of an adult, thereby taking a well-defined place in society, can cause frustration and conflict (Abbink, 2005; Langevang, 2008; Boersch-Supan, 2012). Traditionally, an adult reaches old age when they return to relying on others to look after them. However, intergenerational caring relationships are rapidly changing, for example with many ageing grandparents having become the sole carers of their grandchildren due to the HIV/AIDs pandemic, and thus also experiencing delayed lifecourse transitions. Most elderly urban residents desire to return to 'the village' where they shall resume subsistence farming and rely on their children and other family members for support. At the end of life, it is expected that people will be buried in their natal villages, even if they have been born and lived in nearby cities.

Debates about intergenerational equity in Uganda take place against a national backdrop of a rapidly growing population, local net out-migration, a significant youth bulge and unemployment, corruption, and weak health and education systems. Many urban residents in Uganda were either raised as, or by, subsistence farmers in rural villages rather than in an urban cash economy. Having migrated to cities and towns, they now face the need to diversify into riskier livelihood strategies. This gives rise to concern that younger generations are becoming enslaved by the cash-based economy, reflecting the rising significance of money. Rising inequality in Museveni's generation is perceived to be entrenching short-sightedness, with implications for future generations who will inherit little from their parents and elders, contravening a widely-held belief in parental responsibility to provide a foundation for future generations. Across the continent, urbanization and accompanying economic crises are slowly eroding gerontocratic intergenerational bargains in which established rules on the transfer of resources and responsibilities are breaking down (Collard, 2000; Durham, 2000; Burgess and Burton, 2010; Frederiksen and Munive, 2010; Banks, 2015).

Intergenerational concerns in the UK

Debates about synchronic intergenerational inequality have been especially prominent in UK public discourse in recent years. This discourse is influenced by think tanks such as The

Intergenerational Foundation and the Resolution Foundation, which are proactively lobbying to 'repair' intergenerational relations undermined by the socioeconomic pressures of an ageing society, focusing on issues such as housing, earnings, and tax and benefit policies across generations. These debates typically pit 'Millennials', those born between the early 1980s to late 1990s, against 'Baby Boomers', the post-war generation accused of thriving at younger generations' expense. This debate became particularly high profile following the 2008 financial crisis and recession, with the emergence of publications such as *The pinch: How the Baby Boomers took their children's future – and why they should give it back* (Willetts, 2010) and *Jilted generation: How Britain bankrupted its youth* (Hower and Malik, 2010). This issue has received substantial attention in the national press, with a resurgence following the 2016 referendum on membership of the European Union and numerous commentaries on the generation gap between typically older 'Brexit' voters and younger 'Remainers' (Thompson, 2017). Intergenerational fairness was the focus of a recent public inquiry and report by the House of Commons Work and Pensions Committee (2016), and two recent Chancellors of the ruling Conservative Party have promised voters national Budgets that "put the next generation first" (Osborne, 2016; Hammond, 2017).

The notion of an 'intergenerational social contract' is a recurring theme in this debate. The Work and Pensions Committee's Intergenerational Fairness report notes that '[t]he welfare state has long been underpinned by an implicit social contract between generations' (2016: 3), and in its recommendations talks of 'the intergenerational contract under strain' (46) and 'strengthening the intergenerational contract' (48). Little and Winch (2017) trace the usage of this term back to 18th century conservative philosopher and Whig politician Edmund Burke, who argued in *Reflections on the revolution in France* (Burke and Mitchell, 2009 [1790]: 96t) that: 'Society is indeed a contract. It is a partnership … between those who are living, those who are dead, and those who are to be born.' They contrast Burke's focus on continuity and stability, with the later influence of early 20th century German sociologist Karl Mannheim's theory of generations as distinct birth cohorts sharing a common identity shaped by significant historical, cultural and technological changes that take place in their lifetime (Mannheim, 1952 [1923]). This understanding of generations has similarly entered popular culture, evident in key generational descriptors such as 'Baby Boomers' and 'Digital Natives'.

Until the recent school strikes, climate change had received limited attention within UK public discourse on intergenerational fairness, though it is sometimes subsumed as an issue within the Millennials versus Baby Boomers debate. For example, a 2016 commentary in the national, left-leaning *Guardian* newspaper described climate change as 'intergenerational theft' and, likening it to Brexit, claimed that 'youth will bear the brunt of the poor decisions being made by today's older generations' (Nuccitelli, 2016). Conversely, Millennials are often characterized as more consumption-orientated and thus having a bigger impact on the environment than previous generations. UK environmentalists have drawn on the idea of intergenerational justice to mobilize public action on climate change, typically focusing on family and kinship – the challenges facing 'our children and grandchildren' – as a way to humanise the link between past, present and future generations (Little and Winch, 2017; White, 2017).

A human sense of climate and social change

So far, we have outlined how the convergence of the 'global storm' and 'intergenerational storm' make climate change an intractable problem, and we have considered key arguments for international and intergenerational justice that feature as part of this debate. We have also suggested that long-term or diachronic intergenerational justice considerations are typically less prominent than concerns about synchronic intergenerational equity between people of different generations alive today. In the remainder of this book, we turn our attention to everyday, lived experiences of climate and social change from the perspective of urban residents in Jinja, Nanjing and Sheffield. Through this research, we explore the extent to which environmental justice considerations feature in local ontologies of climate change in different cultural contexts, across cities with disparate levels of resource consumption and exposure to environmental 'bads'.

Recent social science scholarship has drawn attention to the kinds of climate change knowledge that is typically valued in making decisions about our common future, questioning how 'matters of fact about the Earth … become facts that matter for people outside geoscience' (Castree, 2017: 57; see also Gaard, 2015; Mahony and Hulme, 2016). While natural science perspectives predominate in the study of climate change, in recent years there has been a move towards understanding 'the social heart of global environmental change' through approaches that adopt a human-centred analysis (Hackmann et al, 2014). Research increasingly focuses on contextualising climate change alongside other risks – social, economic, political, and environmental – that shape and limit human wellbeing (Agyeman et al, 2002; Pelling and Wisner, 2009; Pelling, 2011; Walker, 2011). Work on the Anthropocene highlights the 'collision' of natural and human history through industrialization, disrupting perceived boundaries between nature and culture (Hamilton et al, 2015). In this context, Hulme (2017) argues that the climate should be understood as both a physical and a cultural phenomenon, and that the 'human sense of climate' – the way people interpret and respond to the climate changing – is important. Barnes and Dove (2015: 5) similarly suggest that 'perceptions of and actions on climate change are shaped by sociocultural relations of power … and by the ways in which people see themselves in the world'. This way of thinking about climate change suggests that the social sciences and humanities can act as 'vital intermediaries' between geoscientific claims about our changing planet and social action (Castree, 2017). Advocates of this approach caution against a deficit-based view of everyday 'unscientific' knowledge, arguing that the social construction of climate change and the ways in which it is conflated with other kinds of knowledge offers valuable insight into possibilities for action (Meze-Hausken, 2004; Brace and Geoghegan, 2010; Eguavoen, 2013).

In this book, we argue that local perceptions of climate change are interwoven with experiences of urban social and cultural transformation, including changing consumption practices, that complicate and cloud its framing as an international and intergenerational injustice. We outline how local narratives of climate change influence perceptions of moral responsibility, sometimes in ways that run counter to international policy rhetoric, drawing on anthropological insights about the relevance of local environmental knowledge for motivating action (Brace and Geoghegan, 2010; Eguavoen, 2013; Rudiak-Gould, 2014). We also consider how intergenerational community-based research can be used to foster dialogue and identify shared sustainability concerns, as a counter to narratives of generational blame for climate change and unsustainable consumption practices.

Overview of this book

This book draws on in-depth social research in Jinja, Nanjing and Sheffield to explore the relationship between local narratives and moral geographies of climate change, sustainable consumption and intergenerational justice. These themes are addressed cross-nationally in each of the following chapters.

First, we consider urban residents' experiences and perceptions of living with a changing climate. In Chapter Three: Local Narratives of Climate Change, we explore the cultural construction of climate, how it is conflated with local weather and high-visibility environmental problems such as air pollution, tree felling, industrial waste and changing land use. We discuss how local explanatory narratives differ in their treatment of climate change as remote in space and time or immediate and locally rooted; and how this affects the extent to which people feel it has a direct impact on their lives. We also consider generational perceptions of climate change, particularly how older urban residents' lived experiences of (de)industrialization shape their hopes or fears in relation to future climate change. In doing so, we consider how residents draw upon geoscientific knowledge and their own experiences of urban infrastructural and environmental change in assessing their (in)vulnerability to climate change.

Explanatory narratives of climate change in turn influence people's views on responsibility for environmental problems. In Chapter Four: Moral Geographies of Climate Change, we consider the effect of contrasting moral readings of climate change employed by urban residents in Jinja, Nanjing and Sheffield. For example, whether people blame 'meta-emitters' in government and industry, themselves, or more amorphous culprits like human nature or God. We argue that this affects the extent to which people are willing to assume responsibility for environmental stewardship; much more so than their relative carbon footprint. In posing the question 'Who is responsible for what?', we explore divergent moral framings of climate change as a problem for *them, there and then* or *us, here and now* and the possibilities of caring at a distance. This includes attention to the intergenerational challenges of climate change vis-à-vis urban residents' perspectives on caring for the future and historical responsibility.

Developing this theme, we consider a particular manifestation of generational blame that connects ideas about (un)sustainable consumption with anxieties about intergenerational value change. In Chapter Five: Intergenerational Perspectives on

Sustainable Consumption, we explore urban residents' narratives of socioeconomic transitions and their perceived impact on contemporary consumption practices. In contrast to the climate change debate that casts young people as innocent victims of past generations' mistakes, we find that younger generations are often blamed for unsustainable consumption practices, with their habits compared unfavourably to their elders. We consider the valorization of resource conservation through narratives of scarcity and frugality, the influence of social conservative moralising discourses such as 'make do and mend' and *qinjian jieyue* (勤俭节约, 'being diligent and thrifty'), and the totemic role of waste in making unsustainable consumption visible. We explore how such ideas reflect concerns about both environmental and social degradation, with some residents attributing climate change to a more general moral decline.

The concluding section turns its attention to the future, specifically how intergenerational community-based research and creative practice can support public engagement, foster intergenerational dialogue and inspire sustainable action. Chapter Six: Imagining Alternative Futures outlines three collaborations that were part of the INTERSECTION programme: a *Write About Time* workshop led by Sheffield poet Helen Mort; intergenerational participatory research to support creative environmental knowledge sharing in Jinja; and a *Sustainability Dancer* sculpture created by Sheffield artist Anthony Bennett. This work offers a lens to reflect on key findings from across the programme, suggesting creative possibilities for engaging with climate change and informing pathways to just sustainability.

THREE

Local Narratives of Climate Change

> 'In my mind, the environment is getting worse, particularly this year. Probably we were not aware of it before, but now it influences our lives. For example, last winter, it was so freezing, that many pipes were broken. This year, there is flooding ... due to heavy rains, suspended services on the subway. This must be related to environmental changes. So most people are concerned about the environment. ... If every year it will be like this, it shows that the climate is getting worse.' (Jin, female, early 40s, Nanjing)

Introduction

This chapter examines how urban residents in Jinja, Nanjing and Sheffield understand and narrate their experiences of 'climate and the ways it might change' (Brace and Geoghegan, 2010: 3). Scholars have highlighted a gulf between expert and lay knowledges of climate change, in particular the contrast between the abstract spatialities and timeframes of global geoscience, and local, grounded experiences of living with a changing climate. This suggests a vital role for social science in understanding the kinds of environmental knowledge that is privileged and valued by ordinary citizens, in particular how 'big' environmental concerns and everyday 'small' environments are connected (Phoenix et al, 2017). Bickerstaff and Walker (2001) argue that public understandings of global environmental change inevitably involve judgement, interpretation and 'sense-making', modes of perception that are inextricably tied to the local context.

When asking people across diverse geographical and cultural contexts about the impact of climate change on their lives, it is important to take into account how the idea of climate – and thus of climate change – may be differently conceptualized (Sheridan, 2012; Eguavoen, 2013; Hulme, 2017).

Across Jinja, Nanjing and Sheffield, our data reveals how residents similarly conflate climate change, weather and more visible, local environmental problems. However, their explanatory narratives differ in their treatment of climate change as remote in space and time or immediate and locally rooted, and this affects the extent to which people perceive climate change has a direct impact on their lives. We also explore generational perceptions of environmental change in the context of (de)industrialization of the urban landscape, contrasting the experience of the former 'Steel City' of Sheffield with narratives of urbanization and environmental decline in Jinja and Nanjing. This allows us to highlight the direct links between urban transformation, infrastructure and land use change, the visibility of pollution and thus the extent to which environmental concerns are prioritized.

"How will I predict the climate when I'm not a scientist?"

Recent research has contrasted geoscientific conceptualizations of climate and more 'local' or 'indigenous' narratives, characterizing the former as 'disembodied global science' and the latter as 'local, embodied knowledge' (Mahony and Hulme, 2016: 2). This work highlights how climate change is typically associated with and dominated by discussions of global atmospheric processes, and thereby disconnected from people's lived experiences (Barnes and Dove, 2015; Castree, 2017). Through such framing, climate change is decoupled from its social context and represented as taking place in an abstract space and time, apart from everyday consumption practices and local perceptions of weather, seasons and landscape (Brace and Geoghegan, 2010; Girvan,

2014). Research has shown that people struggle to interpret the scientific uncertainty and complexity of climate change represented in this way (Lorenzoni et al, 2007). The dominant framing of climate change as 'inaccessibly scientific and technical' (Rudiak-Gould, 2014: 375) has implications for how people engage with it.

This was reflected in our research findings across the three INTERSECTION cites. A majority of residents in Jinja were unfamiliar with scientific discourses of anthropogenic global climate change, but those who had encountered the term (often in the media) felt ill-equipped to engage with what was seen to be an inaccessible expert knowledge reserved for 'scientists' and 'geographers'. In Sheffield and Nanjing the term was more familiar. Nonetheless it was often perceived as a concern for scientists rather than ordinary citizens. When asked how they thought the climate might change in the future, many people were reluctant to offer their own view:

> 'Ha! How will I predict the climate when I'm not a scientist? It's impossible to predict the climate.' (Patience, female, early 60s, Jinja)

> 'It's very difficult to say what it is. Some scientists invent these things. We – we can't do anything about it.' (Humu, female, late 70s, Nanjing)

> 'You've got to listen to them if they go on enough that something is bad for the environment, then it might influence me ... I'm not thinking in terms of the government. They're just ordinary people, aren't they, the politicians, but the scientists come up with various theories, don't they?' (Geoff, male, late 70s, Sheffield)

These responses highlight how the predominant geoscientific framing of climate change can discourage dialogue and knowledge sharing among ordinary citizens. While recent studies suggest a growing public fatigue and scepticism of climate science in Western countries (Capstick et al, 2015), we found that those residents familiar with geoscientific discourses in Jinja and Nanjing tended to defer to scientific expertise, along with many residents in Sheffield. As our data suggests, this has particular implications whereby urban residents can come to expect that the diagnosis of problems and possible solutions to climate change are the responsibility not of themselves, but these same experts.

Cultures of climate

To refer to 'cultures of climate' is to recognize that the idea of the climate changing is more than a description of atmospheric processes. Climate encompasses a whole range of 'deep material and symbolic interactions which occur between weather and culture in places' (Hulme, 2017: 2). For example, the anthropologist Rudiak-Gould (2012, 2014) has observed through his work in the Marshall Islands how the nearest equivalent Marshallese word for climate – *mejatoto* – refers to both physical and cultural surroundings. This means that when Marshall Islanders talk about climate change – *oktak in mejatoto* – they are referring simultaneously to material and social changes in the fabric of their lives.

A key issue raised by anthropological work on climate change is the challenge of translating the term *climate* in cultural contexts where its geoscientific framing is less familiar; a difficulty which raises the question: 'Do we talk about climate or something else?' (Eguavoen, 2013: 7; see also Tschakert, 2007; Slegers, 2008). For those Jinja residents unfamiliar with or reluctant to comment on *climate* and *climate change*, consistent with other anthropological studies we adopted the more inclusive vocabulary of environmental change. In both Luganda and Lusoga (the two dominant languages in Jinja, along with English) the word for environment – *butonde* – refers to the natural and human-made environment surrounding an individual, including not only atmospheric, terrestrial, aquatic and built environments, but also the social and political context and way of living. Questions about environmental change thus invited broad reflections on changes in the weather and seasons, changes in the local urban environment, and how the social fabric of life itself was changing. When Jinja resident Harriet was asked if she had noticed any environmental changes, she described how building and agricultural practices were directly affecting the weather:

> 'People are cutting down trees to build houses and farming, so that affects the environment because we have spent like three months without having any rain and whenever it wants to rain it disappears.' (Harriet, female, late teens, Jinja)

Elderly resident Agatha, reflecting on how the environment had been in the past, said:

> 'In the [19]60s people were more humane, compassionate and friendly and we could talk to each other like a family and even the weather had not changed.' (Agatha, female, early 80s, Jinja)

The binding of social, urban and natural worlds was thus enshrined within the cultural construct of 'environment', with perceived changes in the climate attributed to changing land use and social practices at the local and national scale.

In Nanjing and Sheffield there was more familiarity with geoscientific discourses on climate change (*qìhòu biànhuà* or 气候变化 in translation), with residents often giving 'textbook answers' when asked about environmental problems affecting people alive today:

> 'It might be the global warming. Drought, flooding, there are more and more natural disasters. Coastal areas are losing land.' (Weijia, female, late teens, Nanjing)

> 'Global warming and Antarctica and the polar bears losing their home and the ice is melting and it's because we're using – wasting electricity.' (Emma, female, late teens, Sheffield)

Large-scale survey data suggests that only around 1 per cent of the English public and 7 per cent of the Chinese public has never heard of climate change (DEFRA, 2002, 2007; Wang et al, 2017). On the one hand, this suggests that Nanjing and Sheffield residents are more likely to echo geoscientific perspectives. However, research in Western contexts has shown that despite widespread general awareness of climate change, public understanding of its causes and effects is hugely variable, tied to personal beliefs and values, and mediated by experience (Stamm et al, 2000; Lorenzoni et al, 2007; Whitmarsh, 2009). Furthermore, to align with fieldwork in Jinja, our questioning often employed the terminology of observed or anticipated 'environmental change' (*huánjìng biànhuà* or 环境变化). This invited similarly broad commentary on weather patterns, changing social norms and the built environment alongside various local, national and global environmental issues. This enabled us to explore a more holistic view of how climate change narratives are situated within and interwoven with other aspects of urban material and cultural transformation.

Local narratives of climate change

Discussions of climate change across the three INTERSECTION cities frequently elicited concerns about local issues such as tree felling, air and water quality and littering. This extends and supports previous research which has shown that people conflate climate change, weather and other environmental issues at various scales, such as the hole in the ozone layer, local air quality and waste (Lorenzoni et al, 2007; Hulme, 2017). For example, when Hui, a Nanjing resident, was asked whether she and her family had ever discussed climate change, she replied:

> 'Oh, we discussed when the smog got a bit heavy; we told each other to be careful when we went out, to wear masks. Old people shouldn't go out. Then we discussed that we were very careful, all of us didn't drive to work, and we all felt that we did very well.' (Hui, female, late 40s, Nanjing)

In Jinja, when Grace was asked whether the way people live has had an impact on the climate, she replied:

> 'I think so because when you look at the environment – do you know buvera [plastic bags]? We get buvera and throw them in the water. Like of recent, just down there, the swamp there had clean water but now people have made it their dustbin. It's now a collection of wastes!' (Grace, female, early 20s, Jinja)

When scaled up issues such as deforestation, air pollution and waste do contribute to climate change. However, urban residents did not typically frame these issues in the context of global trends in environmental degradation or their contribution to greenhouse gas emissions. Rather, they were primarily of concern because of their perceived impact on local environmental quality, as Cathy in Sheffield noted:

> 'The whole of Sheffield would benefit if everybody just did their bit … like companies sticking to within the guidelines that they set for pollution and pollution of air and rivers,

> and for people just not creating any litter. I don't think people take as much care in an industrial city as they do probably in other parts of the country.' (Cathy, female, early 50s, Sheffield)

In these examples, lived experiences of environmental pollutants are represented as both a drawback of urban living and a localised threat to human health and quality of life. In the following section we explore residents' shared concerns about tree felling and its impact on the urban environment, before turning to the intricate relationship between weather and climate in local narratives.

"Everyone wants to live in a city full of green trees"

Tree felling is a prominent example of an issue conflated with climate change that had strong resonances across all three cities, where residents' concerns had little to do with how it might contribute to the greenhouse effect. In Jinja, it was perceived to have a negative impact on soil quality, rains, storm protection, and the sustainability of food, timber and fuel supply:

> 'There has been a lot of deforestation because of charcoal [production], for burning. Even building our houses, they waste a lot of timber and those people do not plant more trees when then are cutting, they just cut and leave. So now deforestation affects the weather.' (Maria, female, late 50s, Jinja)

> 'People are just keeping on building and building, then which means there will be no wind breakers which trees can do – so when the rain comes, all the heavy rains come, they easily break the house that is next to your house and then that house can fall on yours and then the other one like that.' (Francis, male, late teens, Jinja)

Deforestation and affordable energy are interlinked sustainability challenges in Uganda, with biomass being the most important source of energy for the majority of the population, accounting for 94 per cent of total energy consumption (MEMD, 2013: 101). Four million metric tons of wood are consumed annually to meet the demand for charcoal (Ferguson, 2012: 2), which provides fuel for household consumption. The rising cost of charcoal is a major expenditure for the urban poor, giving rise to informal and illegal tree cutting in neighbourhoods where residents are struggling to make ends meet (McQuaid et al, 2017).

In Sheffield, tree cutting was a high profile local political issue at the time of our fieldwork due to a decision by Sheffield City Council to fell and replace thousands of trees assessed as dead, dying or dangerous. Many residents opposed this scheme and extolled the benefits of trees:

> 'We need to keep planting trees; Sheffield is very green and we're very proud of that and we shouldn't lose that. We need to stop urbanising green spaces in general with concrete, and because of the flooding that we had.' (Ann, female, early 50s, Sheffield)

> 'Trees we've had for millennia and so, for me, it's the sustainability of those trees and the fact that they protect me and this house from the road and a little bit from the wind … They belong. Because these were planted I don't know how many years ago, they've probably gone through perhaps two, maybe three families or inhabitants of this house.' (Lily, female, late 60s, Sheffield)

In Nanjing, residents participating in an intergenerational theatre workshop developed a short improvised scene where an 'angel' tree fought the 'devils' of waste and air pollution. Interviewees similarly emphasized the importance of trees in greening the urban environment, making it more 'natural' and pleasant:

> 'In terms of natural environment, everyone wants to live in a city full of green trees, and no one wants to live in a dirty filthy place.' (Lingping, female, early 20s, Nanjing)

For urban residents, trees occupied a totemic role in discussions of climate change, the environment and sustainability. Previous research has highlighted that, as well as aesthetic and practical benefits, trees have a particular psychological value to urbanites as a source of connection with 'everyday nature' and as a significant contributor to health and wellbeing (Dwyer et al, 1991; Lohr et al, 2004). The felt presence of trees as 'angels' and protectors from storms, floods and pollution illustrates how local perspectives on climate change include normative assumptions about what 'belongs' in the urban landscape, with the idealized trope of the green city evoking ancestral heritage and emotional connection to place alongside practical concerns. In the next section, we explore how this also extends to idealized accounts of remembered weather.

"People expect the rain to rain": weather as (dis)proof of climate change

Non-expert perceptions of climate change are refracted through the lens of local weather, seasons and expectations of climate normality (Eguavoen, 2013; Capstick et al, 2015; Hulme, 2017). Climate refers to long-term atmospheric trends and the expected seasonal 'behaviour' of the weather in different regions (Hulme et al, 2009), but for many the distinction between weather and climate is confusing. This was acknowledged by Sheffield resident Joe:

> 'You could say there's been extreme weather, but then I think it's just because we think it is, so when it's heavy snowfall, is that because it's environmental or is that just heavy snowfall? … I think that's a lot of the problem

> with climate change, that there isn't anything obvious that people can show. If a summer is one degree hotter than it was ten years ago, people in general aren't going to notice that, even though it's significant.' (Joe, male, late teens, Sheffield)

In other words, extremes and anomalies in daily weather are what people tend to recall when asked to think about climate or environmental change, irrespective of whether these weather events are a reliable indicator of long-term trends (Hulme et al, 2009; Li et al, 2011). Indeed, many residents in Nanjing and Sheffield offered contradictory observations of how the local weather had changed, for example whether Nanjing winters are becoming warmer, colder or more or less the same as always:

> 'I can see it [climate change] is happening from little signs. Winter is warmer than before.' (Daowei, male, late 40s, Nanjing)

> 'Now you can barely see such snow as there used to be. We go out to take photos even after a light snow. It seems like the snow has become a rare thing here.' (Zhenzhen, male, late 70s, Nanjing)

> 'This winter it was minus 9 degrees. It had not happened before.' (He, female, late 70s, Nanjing)

> 'I've heard of greenhouse effects which would – but I find that it is not obvious; it is still very cold in winter. Some theories are not very true.' (Qun, male, late 40s, Nanjing)

Likewise, Sheffield residents offered multiple perspectives on changes in snowfall, rains and summer temperatures in living memory:

> 'I remember as a child winter and summer were polar opposites. We had very cold winters with snow up to your waist, and nice, warm, extended periods of fine weather during the summer. Whereas now, we – things merge into each other.' (Stu, male, early 40s, Sheffield)

> 'I feel like it's been a lot warmer and sunnier this September than I can ever remember before, so that's the only way that I feel I've really experienced it from just different weather, different weather at Christmas. It felt really strange last year, it being so warm on Christmas Day.' (Hailey, female, early 20s, Sheffield)

> 'We've had some bad, some bad rains. I think a lot of it is to do with that El Nino phenomenon. I don't – this global warming thing doesn't add up. It's not true.' (Lester, male, late 60s, Sheffield)

This illustrates that, in Nanjing and Sheffield, urban residents' interpretation of local weather as proof (or disproof) of climate change was inconsistent. Residents tended to focus on one or two standout – usually quite recent – periods of unexpected weather in contrast with a nostalgia-tinted past (Harley, 2003). In both urban and rural environments, people's first and foremost encounter with climate knowledge is through the weather they experience day to day, and thus '[f]or most people the idea of climate becomes reified through a rather unstructured assemblage of remembered weather' (Hulme, 2017: 30). Idealized memories of past weather such as deeper winter snows and brighter summers shape expectations of how weather *should* behave and make the prospect of novel climates difficult to imagine.

Some residents, like Joe, recognized the difficulty of disentangling 'climate' from 'weather'. In a similar vein, Zaafir reflected on the limitations of his ability to directly perceive climate change:

> 'I've experienced in this country some very hot summers, very few really cold winters, but in the scheme of time and weather maps chronologically that's not an unusual thing. In a 50-year cycle that is not an unusual thing. However, some of the data is saying it's warming. But we've had September days like this before. It's not the first time. I've also had snow in June, so I've seen that. It's very difficult because you don't – we're not really seeing it at a daily level.' (Zaafir, male, late 50s, Sheffield)

This presents a challenge if, on the one hand, geoscientific knowledge of climate change is too far removed from everyday life, while on the other hand lived experiences of weather and climate are an unreliable barometer for understanding global environmental change. Rudiak-Gould (2013) suggests that the extent to which climate change is 'visible' is simultaneously an empirical question about how much humans have altered the planet, an instrumental question about how best to communicate the climate change threat to the public, and a political question about the authority of scientific and citizen knowledge respectively. For better or worse, people tend to trust their personal experiences of the environment (Bickerstaff and Walker, 2001). Some scholars caution that this makes public opinion on climate change 'as mercurial as the weather' (Li et al, 2011: 5), with local weather observations cited by the concerned and sceptical alike to support their arguments.

In Jinja, local narratives of weather and climate change – while similar in their evocation of a nostalgic past of remembered weather – were more consistent, reflecting concerns about two periods of drought in 2015 and 2016 that coincided with our fieldwork. In both years, rains usually anticipated for arrival in February did not appear until late April. Droughts are on the increase in Uganda with negative consequences for agricultural production, food security, water supply and community livelihoods (Akwango et al, 2016). The problem is likely to intensify: climate projections indicate that by the year 2050, most of Uganda will experience a rise in temperature of between 2°C and 2.5°C in the warmest months (Bashaasha et al, 2013). In 2015 and 2016 residents also observed more intense rainfall, especially in the second rains towards the end of the year, causing flooding and erosion. Climate change was thus primarily interpreted as a local phenomenon of increasingly erratic rainfall distribution. Christine, who engaged in peri-urban agriculture selling maize and bananas, explained:

> 'It has been very dry so it has changed a lot … Long ago there used to be a lot of food and the rains would come and we would plant everything but now we do not have food, we just buy … The rains when they come these days there is a lot of storm that instead of coming the good rains where you would plant things and harvest them, it comes with a lot of wind; all the banana trees they fall down, the houses are unroofed.' (Christine, female, late 40s, Jinja)

In Uganda, annual rainfall is less critical to farmers' production than distribution through a season, the way rain falls during rainfall events, and the impacts of increased temperature on soil moisture (Osbahr et al, 2011: 294; see also Mukiibi, 2001). With 80 per cent of Uganda's population dependent on rain-fed agriculture, drought represented a significant problem for Jinja's residents. Their narratives reflected worries about the unpredictability of when the rainy seasons would begin, how long they would last and their perceived reduction in number:

> 'Like in this season people expect the rain to rain like in August but it is just sunny so they expected to plant crops and some planted in the sunny season and the plants did not come out well. It rains just once in a while which was not the case before … The families also lack what to eat and if you find a big family, it cannot afford to buy

> enough food for the family member[s].' (Doreen, female, mid-30s, Jinja)

A number of residents described how, when the rains did arrive, they did so in the form of destructive storms, wreaking havoc on crops, houses and roads. With poorly maintained drainage systems and roads, inadequate sanitation, and overcrowded ill-maintained houses, extreme weather events could have devastating impacts, particularly on poorer communities.

Jinja's residents' accounts of how the local weather has changed tell a more consistent story than those in Sheffield and Nanjing, and one that appears to confirm expectations of how climate change will affect urban populations in sub-Saharan Africa. However, this may to some extent be an artifice of our fieldwork happening to coincide with a period of drought. This is a good illustration of the drawbacks and benefits of making climate change 'visible' that Rudiak-Gould (2013) outlines. On the one hand, links between local weather and global climate can be 'empirically dubious' and it is problematic to attribute straightforward causality. On the other, Jinja's residents themselves are not making such claims, rather they are articulating their lived experience of vulnerability to climate shocks in a particular locality. As Rudiak-Gould (2013:127) argues, the 'most fundamental and consequential effect' of paying attention to these narratives, particularly in communities that have strong intergenerational attachments to place, 'is to give nonscientists the right to speak about climate change' and to mobilize local ecological knowledge towards climate adaptation (see also Vedwan, 2006).

"We're in the wrong place": perceived (in)vulnerability to climate change

Despite some commonalities among the three cities in the conflation of climate, weather and urban environmental issues, there was also an important difference in how climate change was conceptualized and thus the extent to which urban residents perceived it to have a direct impact on their lives. The difference is most starkly illustrated by our survey, which found that far fewer people in Sheffield said that they think often about climate change (Figure 3.1) or that it has a major impact on them personally (Figure 3.2). Compared to Sheffield, over twice as many residents in Nanjing and Jinja said that they 'often' think about climate change. More than six times as many residents in Nanjing and almost ten times as many in Jinja said that climate change has a major impact on them personally. These findings could be interpreted as reflecting different levels of regional vulnerability and exposure to climate change. However, we suggest that they also reflect differing local interpretations of the meaning of climate change: specifically, whether climate change is predominantly perceived as a global phenomenon that is remote in space and time, or one immediately rooted in the local environment.

For Jinja's residents, problems exacerbated by drought and irregular rainfall such as the availability and price of food and the failure of their crops were a profound source of worry:

> 'It has affected [us] now, because the little people have planted it has dried off. Now here we lack food, you go to the market there is no food: no rain, no food. Then the price [of food] has gone very high, you cannot be able to afford to buy. It has affected us! Hunger.' (Hilary, female, mid-30s, Jinja)

It is little surprise in this context that residents perceived a direct link between climate change and their everyday lives, irrespective of their familiarity with discourses of global anthropogenic climate change. As we explore in Chapter Four, many of Jinja's residents believed that the changing weather was caused by local practices such as tree felling, construction and sociocultural

Figure 3.1: How often have you thought about climate change?

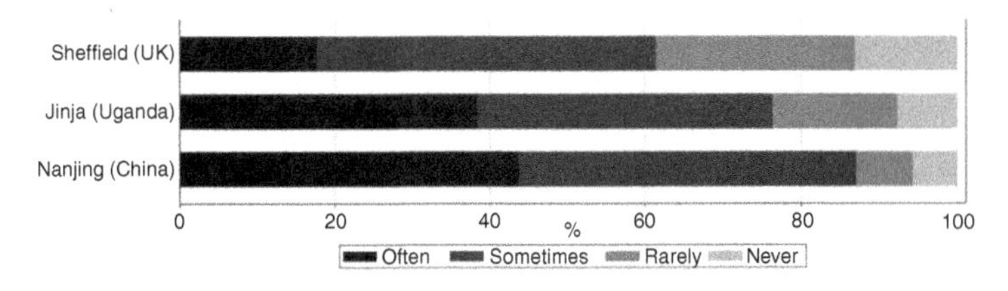

Figure 3.2: To what extent is climate change having an impact on you personally?

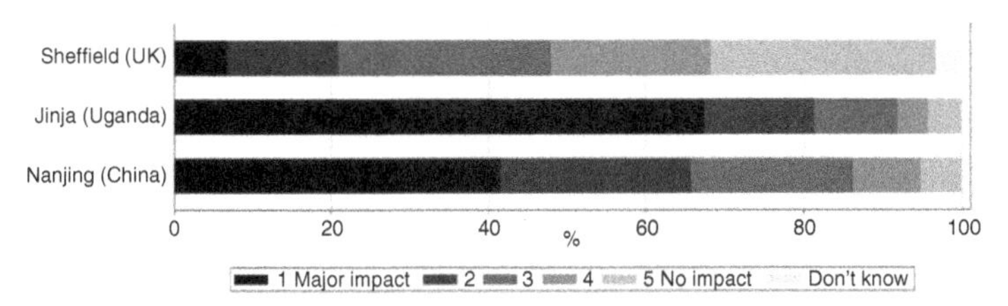

Note: 'Don't know' response category was only available in Sheffield.

changes. Their conceptualization of climate change and its impact thus centred on their local community, as David, a local journalist, explained:

> 'I don't think the environment means the ozone layer straight away, but the environment around us – according to me, the environment around us and how we treat it contributes to the bigger picture whether you're destroying or you're maintaining which transmits up to the bigger picture of the ozone layer.' (David, male, early 30s, Jinja)

This suggests a bottom-up view of climate change as rooted in place and practice, with local actions having both local and global repercussions for the environment.

In contrast, most residents in Sheffield described climate change as affecting other places ("obviously it's much worse across the world just from reading in the news") and/or future generations ("you're talking about some abstract thing that's not happened yet"). This is consistent with a tendency among the UK public to perceive climate change as a distant problem (Lorenzoni et al, 2007; Marshall, 2014):

> 'We're in the wrong place for it to be affecting us. It's places like the Marshall Islands, which are about to go under. I mean, it's other parts of the world where it's noticeable at the minute. It'll become more noticeable here but not for a long time, and also not as drastically probably.' (Rose, female, early 70s, Sheffield)

Interestingly, very few Sheffield residents discussed the 2007 floods – the city's standout recent experience of extreme weather – as possible evidence of climate change, or in the context of future flood risk. One three-generation family in our study had suffered major property damage and been temporarily rehomed for several months due to this flood. However, this was only revealed by the granddaughter in the last of the family interviews. Neither her mother nor grandmother, who live in adjacent houses, mentioned their experience of flooding when asked if they had noticed any signs of climate change:

> 'My aunt who lives next door with my nana [grandmother] has lived in this row of houses all her life. Her parents owned that house before her. In her lifetime the water has never risen to floor level. It's come up in the cellars but it's never actually flooded the properties. Then in 2007 it happened twice in ten days … Then since we've almost been flooded multiple times and the houses down the side street get flooded a lot more. So it's obvious that climate change is happening.' (Debbie, female, late 20s, Sheffield)

Though local weather was frequently cited as anecdotal proof or disproof of climate change, there was also a reluctance among Sheffield residents – of which Debbie's comment is the exception – to suggest that it has negative domestic consequences. Rather, climate change was more typically cast as a disaster that visits poorer nations:

> 'Climate change. Maybe not so much in this country but in other countries, certainly. When you're looking at drought and flash flooding – I suppose we have had flash flooding, haven't we, here? We had our big flood in Sheffield a few years ago. So yeah, I don't really remember any – I remember the Ethiopian Biafran famine[1] when I was a child, I remember seeing photos and news issues on that on the television.' (Amy, female, early 50s, female, Sheffield)

Amy's comment is revealing of the ways in which Sheffield residents distanced themselves from climate change, imagining it as 'conceptually as well as physically remote and exotic' (Phoenix et al, 2017: 128) through dramatic news headlines and impressions of wholesale ecological disaster in industrializing and poor nations. Amy's recollection of the "big flood" and simultaneous insistence that she doesn't really remember any impacts of climate change reflects a belief in climate invulnerability in the everyday lives of Sheffield residents. In contrast, the particular vulnerabilities of the urban poor in low and middle income countries to climate change arise not only from extreme weather, but as a result of the urban infrastructure's capacity to cope with climate shocks (Dodman and Satterthwaite, 2009).

[1] The famine that Amy refers to actually occurred in West Africa and was a human-induced food shortage triggered by a blockade during the Nigerian-Biafran war (Norman and Ueda, 2017).

In Nanjing, climate change was conceptualized both in terms of the deterioration of the local environment and global atmospheric change. Hang was asked whether she is concerned about climate change and replied "Absolutely, our family is very concerned about climate change." When asked why, she elaborated:

> 'Well, because if the smog is serious, or it is foggy, we will not be able to go out. If I just go out to buy groceries, then we would not go out on that day. Other than that, we are now really worried about, you know, the current weather, weather change. So many days a year have bad weathers. It makes me quite anxious, really quite anxious.' (Hang, female, early 70s, Nanjing)

Nanjing residents' accounts of climate change often oscillated between changes they had observed in their city and changes they had heard of in other parts of the world:

> 'I have noticed some change especially in recent years as we have experienced some extremely hot and cold weather. In some regions such as the melting of ice in the Arctic and the Antarctic and the ozone layer problem, there are many reports about these problems.' (Niu, female, early 20s, Nanjing)

Some residents' main priority was the deteriorating quality of the local environment. Hui, in her late 40s, said "The environment seems to get really worse. I think that we should pay attention to it immediately, otherwise it will get worse and worse", but also admitted: "I don't care much about the fact that the ice in the North Pole is melting quickly." Others however argued that global climate change does concern them. Jiang, in his late 70s, also discussed the icecaps melting and said that while "there is no way you can feel these changes", they have an impact because "global warming can have an influence on human's life expectancy and mental status." Thus, for people in Nanjing, climate change tended not to have the same remoteness attached to it as it did in Sheffield, nor the same exclusively local explanation as in Jinja, reflecting a heightened awareness of anthropogenic climate change among the general public (Wang et al, 2017) alongside particular concerns around the environmental impact of China's recent development, which we explore in the next section.

Generational narratives of environmental change

Eguavoen (2013) suggests that there is a prevalent assumption in climate change adaptation literature that elder generations will be more knowledgeable about climate change because they have had more time for observation. While it is true that climate change requires a long-term view, there are two problems with this assumption. The first is the empirical question of climate change visibility in everyday life, which Eguavoen terms the 'low local perceptibility of global change' (2013:7). The second, related problem is how climate change narratives are interwoven with other aspects of social change. When elder generations are cast as being best placed to bear witness to climate change, Eguavoen argues, 'the abundantly obvious fact that they have also witnessed enormous transformation in terms of technology, political and economic conditions, as well as demography, is often neglected' (2013:12). This is well illustrated by our interview data from Sheffield, where older people's narratives of environmental change were principally influenced by their lived experiences of deindustrialisation.

"I've not seen a peasouper since I was a teenager"

When asked if they had noticed environmental changes in their lifetime, older Sheffield residents said there have been changes for the better as a result of deindustrialisation and improved environmental regulation. Nancy, in her late 70s, recollected "I've not seen a peasouper since I was a teenager", describing a thick fog caused by air pollution from the factories. These narratives often pivoted around the Clean Air Act of 1956 as a turning point:

> 'All the buildings in Sheffield were black … and that was because the atmosphere in Sheffield was black … We lived in it, we lived with that, and we didn't really know that it was so polluted except that when it wasn't being polluted it was a different atmosphere. So I think most of that's gone, because of modern thinking. Also the rivers, the river through Sheffield was always horrible and polluted and the canal was polluted … Now it's fairly clean all the way down to where it meets the Humber.' (Geoff, male, late 70s, Sheffield)

> 'When I first started work when I was 16, the bus used to come down through Five Arches opposite the Wednesday [football] ground and as you look down the valley, you couldn't see more than 100 yards in front of you because of the smoke from the factories. The only time you saw daylight like this was two weeks in the summer when all the steelworks closed down and went on holiday. Otherwise it was – well it was smoke all over the place.' (Bob, male, late 70s, Sheffield)

In addition to the cleaner air and water, older residents discussed Sheffield's abundance of green spaces and evidence of biodiversity such as fish returning to the inner-city rivers, evoking an idealized rather than remembered past. Younger residents also referred to Sheffield's "dingy" and "grim" industrial past in contrast to their own experience of the urban environment:

> 'When I think of my granddad's generation, they lived in a time where Sheffield will have been under a blanket of coal burning from all the factories. It must have been quite grim sometimes. Now we've gone more into the service industries in Sheffield, so you haven't got a lot of that anymore. To my granddad especially everything must seem a lot cleaner now.' (Grace, female, early 30s, Sheffield)

While our research suggests that Sheffield's residents could differentiate between positive environmental changes in their city region and global climate change, it also shows that people privilege immediate and observable changes that affect their daily quality of life over less visible global climatic trends (Bickerstaff and Walker, 2001). Demonstrable improvement and greening of the urban locality thus reinforced their view that climate change is a problem for people elsewhere, primarily in more recently industrialized nations:

> 'In the 1850s we were the workshop of the world. We manufactured the most. We were probably providing the most pollution at that time. We're in a post-industrial society and China is just moving to an industrial society so they're only doing what we did … I see China as such an immense place, I can't imagine what the effects of their pollution will be and the same with India. It's a huge amount of industrialisation.' (Stacey, female, late 50s, Sheffield)

Reflecting on Sheffield's transition in living memory to a less visibly polluted city, many residents had an optimistic view of human capabilities to deal with environmental problems, suggesting that other cities will follow a similar roadmap. This is reflective of a predominant 'ecological modernization' framing of sustainable development in Western contexts that equates economic growth with development, suggesting that advanced industrialized nations have been most successful at development, and that win-win solutions to climate change can be found through 'better' modernization (Lewis, 2000; Barry, 2003; Bailey and Wilson, 2009).

"The planet we took from our parents' generation featured green mountains"

In this respect, Sheffield residents' narratives of environmental change were in stark contrast to those in Jinja and Nanjing, who perceived their environment as deteriorating:

> 'This one of these days is worse than those days. Me I don't know why it comes also. But those days weather was very good, climate was very good, but these days it is changed from those days.' (Ajani, male, late 70s, Jinja)

> 'The planet we took from our parents' generation featured green mountains and clean water, but now everything has changed.' (Feiyi, female, late 20s, Nanjing)

For Nanjing residents, smog was the most commonly cited change, with many expressing fears for their own health or that of younger and older members of their family. This issue is often conflated with climate change, with 95 per cent of respondents to a recent national survey saying that climate change will increase the occurrence of air pollution (Wang et al, 2017):

> 'When I was a child, I didn't know what smog was or rather I'd rarely seen such a phenomenon. But now you see it more and more often … now, you see, the sky is always grey.' (Daowei, male, late 40s, Nanjing)

In Jinja, the recent droughts and food shortages were blamed on urban development and tree felling, for reducing the availability of productive land and contributing to soil degradation:

> 'People have constructed. All the places there [had] bushes around but people have constructed. We used to plant cassava around there and we would get food. But everywhere houses have been mushrooming like that.' (Christine, female, late 40s, Jinja)

In both cities, while recognizing the necessity of industrial development, residents were concerned about its impact on soil, air and water quality, and – as we discuss further in Chapter Four – perceived that the quality of their local environment was secondary to growth:

> 'The factories, now the leather tanning factory down that side, it pollutes the air so much, but also there's something more dangerous it does … they were dumping their waste and they were channelling it to the lake or to the river … and that concerns we because we may not see the effects now but in the future we might start facing it.' (David, male, early 30s, Jinja)

> 'I remember when I had just come to Walukuba[2], there were few houses and then also there was a drainage system where the water could easily pass and it was up there and right now when you go there, it is no longer there because they have built factories around there and no one is caring about it and I think in the future, the water will be so little.' (Francis, male, late teens, Jinja)

> 'The ecological environment has changed a lot, in every aspect. During the process of urbanization, industrial development and also road and mine development, we've been experiencing … a deterioration of [the] ecological

2 Walukuba is a suburb of Jinja.

> environment … like water loss and soil erosion as well as weathering.' (Fu, male, early 50s, Nanjing)

Though the extent of industrialization and urbanization is not equivalent between the two cities, Jinja is currently experiencing a modest industrial renaissance with the opening of new factories (such as those manufacturing soap, plastic goods, textiles, alcohol and steel) and a slow growing stream of foreign investment, mostly from China and India. Industrial development particularly impacts upon the urban poor living in former workers' estates with overcrowded and deteriorating housing and infrastructure rising up from the wetlands of the lakeshore. These narratives from Jinja and Nanjing highlight how urban infrastructures are integral to the cultural construction of 'environment'. They demonstrate the connection between lived experiences of urban vulnerabilities, environmental health hazards and perceptions of climate change. These residents' more palpable concerns about climate change cannot be entirely disentangled from their more recent experiences of rapid urban development and the visible pollution it has brought with it, in the same way that Sheffield residents' complacency about climate change is somewhat attributable to their belief that they live in a comparatively post-industrial, cleaner and greener city than that of the past.

Chapter summary

It is important to avoid 'climate reductionism' (Hulme, 2011) and thus pay attention to everyday, grounded environmental knowledge. In this chapter, we have explored how climate change is differently conceptualized across three cities, highlighting 'the localisation of people's understandings within the immediate physical, social and cultural landscape' (Bickerstaff and Walker, 2001: 133). Our research illustrates how local narratives of climate change are contingent upon urban infrastructural and environmental change, and clouded by lived experiences of weather and seasons. It also reveals how the framing of climate change as a global or local phenomenon has implications for public engagement. Jinja's residents largely made sense of a changing climate and recent drought through their own narrative frames, rather than through geoscientific discourses of global climate change. As such, they were more inclined to believe that climate change has a direct impact on their lives and vice versa through the attribution of blame to local land use and social practices. In contrast, Sheffield's residents tended to think themselves invulnerable to climate change, generally accepting the gist of global climate projections but casting it as more of a problem for industrializing nations and future generations. In Nanjing, the pace of urban and industrial development in recent decades fuelled residents' concern with local and global climate change and environmental health hazards. In the next chapter, we discuss how these very different local explanatory narratives of climate change proximity and remoteness influence residents' beliefs about culpability and responsibility for environmental problems.

Photo 1: Nanjing residents discuss their consumer habits and environmental concerns

Photo 2: Nanjing residents workshopping ideas for a play called 'Supershop'

Photo 3: Jinja residents perform the Kingfisher play as part of a waste action intervention day

Photo 4: Older residents perform 'We Are the Foundations' at a workshop with Jinja Municipal Council

Photo 5: Intergenerational discussion at a workshop in Sheffield

Photo 6: A Sheffield resident who took part in the *Write About Time* workshop reads the finished pieces

Photo 7: Head of the *Sustainability Dancer* sculpture

Photo 8: Detail on the *Sustainability Dancer* sculpture

FOUR

Moral Geographies of Climate Change

> 'People who come from where you come from are the ones causing it, I hear! But to some extent it is not as it was because initially we had a lot of trees and swamps and these could help to maintain the climate but nowadays people are building in a swampy place, a lot of trees are being cut down because of charcoal and firewood. So, this also has an impact on that. The forests that were so big are no longer big because people are cutting the forest to build houses, the factories are many.' (Angela, female, early 30s, Jinja)

Introduction

Local ontologies of climate change – how people see the relationship between themselves, others and the changing climate – play an important role in mitigation, adaptation and willingness to assume responsibility for environmental stewardship. In this chapter, we examine contrasting moral readings of climate change, considering how urban residents' perceptions of climate change in terms of its remoteness or proximity are interwoven with beliefs about the blameworthiness of local and global actors, individual efficacy and personal agency (Jamieson, 2010). The extent to which people view climate change as a distant problem for *them, there and then* (Marshall, 2014) has significant implications for public engagement. We consider both spatial and temporal framings of responsibility for climate change, including 'industrial' and 'universal' blame stances (Rudiak-Gould, 2014), how notions of caring for the future vis-à-vis direct descendants contribute to a relatively short-term 'generational timescape' (White, 2017), how livelihood insecurity risks 'unimagining' the future, and ambivalence towards historical responsibility.

Who is responsible for what? Climate change blame narratives

The term 'anthropogenic climate change' casts human beings as culpable agents who have caused climate change, but climate change blame narratives can range from everyone to no one, variously placing responsibility with fossil fuel companies, affluent consumers, local and national government officials, a provoked God, capitalism, technological hubris, and past or present generations (Rudiak-Gould, 2014; Hulme, 2017). Climate change complicates conventional understandings of moral responsibility, because 'it is not a matter of a clearly identifiable individual acting intentionally so as to inflict an identifiable harm on another identifiable individual, closely related in time and space' (Jamieson, 2010: 437). Instead, it involves vast networks of actors whose influence extends all over the world and far into the future. Persson and Savulescu (2012) argue that human morality, which evolved for life in small societies, is not well equipped to deal with these increased powers of action. Given these challenges, and the different ways that residents in Jinja, Nanjing and Sheffield conceptualized climate change as a global or local phenomenon, we were interested to learn more about who they blame for unsustainable practices and who they hold responsible for acting to address them.

Climate change blame, causality and responsibility for environmental stewardship were key themes addressed in our survey and discussions with urban residents across Jinja, Nanjing and Sheffield, which revealed both similarities and differences in local perceptions. As Figure 4.1 illustrates, most survey respondents thought that human actions contribute to climate change, with Jinja's residents in particular believing this to be a major cause. Industry was commonly identified as a major contributor, and most residents thought that natural changes might play some role. In Nanjing and Sheffield, residents tended to dismiss the idea that 'God or other spiritual causes' might play a role in climate change or said they didn't know. This divided opinion in Jinja: around half of survey respondents believed this to be a contributing factor, and one in five said this is a major cause of climate change. Figure 4.2 shows that across all three cities, residents broadly agreed that certain key actors, such as their national governments, the United Nations, scientists and industry have a major role to play in protecting the environment.

The shared emphasis on national governments, industry and scientists is unsurprising given the predominant global framing of climate change as a technical problem requiring particular forms of environmental governance and expertise (Barnes and Dove, 2015; Mahony and Hulme, 2016; Castree, 2017). However, Figure 4.2 also shows that there were key differences between the three urban areas. Consistent with their more widespread belief in religious explanations for climate change, more residents in Jinja envisaged a major role for religious institutions (though a sizeable minority in Nanjing and Sheffield also said that such institutions have a major role to play). Sheffield residents were more cautious in assigning 'a major role' to any of the identified key actors. The most striking difference was in Jinja, where a majority of residents emphasized the major role played by themselves and their community, in contrast to Nanjing and Sheffield residents' attribution of responsibility primarily to governments and industry. At first reading, this seems counterintuitive given the extent to which the carbon footprint of the average Chinese or UK citizen dwarfs that of the average Ugandan (World Bank, 2014; International Energy

Figure 4.1: Which of the following things do you think cause climate change?

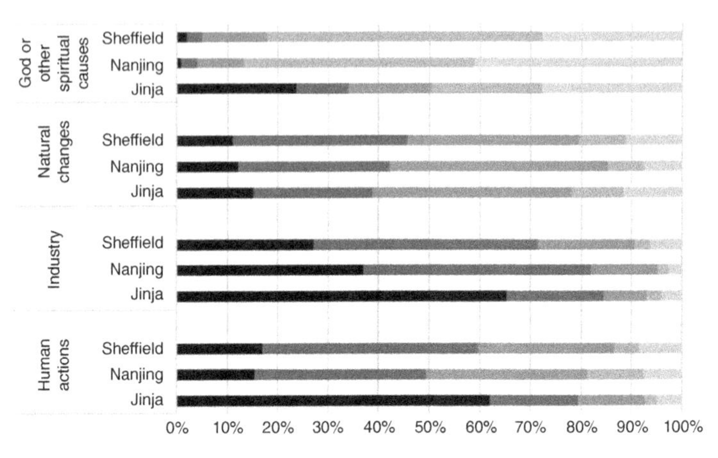

Note: For this question and others in this chapter, an additional 'Don't know' response category was available in Sheffield only, so the Sheffield percentages in Figures 4.2, 4.3 and 4.4 do not add up to 100.

Agency, 2016). This begs the question why urban residents whose lifestyles are more implicated in climate change are less willing to assume responsibility for environmental stewardship, than others whose impact is negligible.

Further analysis of our qualitative data suggests that underlying the headline similarities and differences in our survey findings are contrasting moral readings of climate change. Rudiak-Gould (2014: 366) identifies two particular stances, which he terms 'industrial blame' and 'universal blame', that are helpful for interpreting this data. Industrial blame encapsulates the *climate change as an injustice* argument outlined in Chapter Two, that some people are more responsible for climate change than others and that they ought to be held accountable. Specifically, this stance focuses on 'Western, Northern, developed or industrial countries and citizens', capitalism, corporations and consumer

Figure 4.2: To what extent do these organizations and individuals have a role to play in protecting the environment?

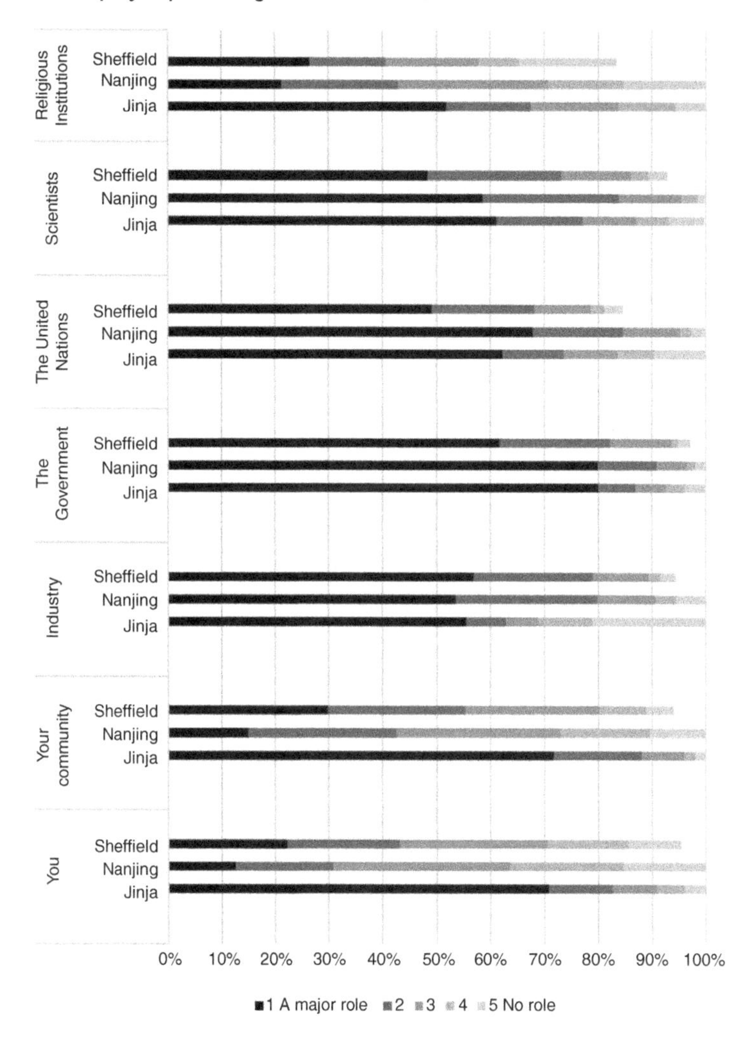

culture as the root of the problem. In contrast, universal blame stances frame climate change as a result of humanity's short-sightedness and self-destruction, focusing on shared culpability and collective responsibility. This stance is common among Northern governments and agencies who have a vested interest in depoliticized readings of climate change, but also among indigenous communities on the front lines of climate change, whom the industrial blame stance casts as victims of injustice. In the following sections, we explore how Jinja's residents held themselves responsible for climate change, in contrast to the prevalence of industrial blame in Nanjing and Sheffield.

"We people are the people who are changing the world"

The quotation from Jinja resident Angela at the opening of this chapter is exemplary of a universal blame stance, expressing her belief in shared culpability for the state of the environment. In this case, Angela appears to be familiar with discourses of Northern or industrial blame, first saying to her British interviewer that she has heard that "people who come from where you come from" are the ones causing climate change. She then goes on to identify local contributory factors such as urban development in wetlands, deforestation, domestic energy use, and factory construction. Many residents' narratives of climate change in Jinja shared Angela's concern with these destructive practices. These were interpreted as signs that the balance of human-environment relations in Jinja has changed, that it is "not as it was" and "this also has an impact on that [climate change]". In some ways, Angela's comment is atypical in blaming foreign actors alongside local practices. More generally, while showing some evidence of rhetorics of universal blame, narratives in Jinja were more suggestive of a third position that emphasized local causality. Climate change was understood as humanity's self-destruction, but a local humanity rather than a global one.

When asked what was causing changes to the climate such as the droughts and rainstorms described in Chapter Three, a very small number of Jinja's residents blamed foreign influences, while considerable numbers invoked changing urban lifestyles and the will of God in some form. Religious explanations were not necessarily fatalistic, nor did they absolve humans of responsibility (Dieter and Bergmann, 2011; Eguavoen, 2013; Hulme, 2017). Whether or not responses had religious overtones, climate change was overwhelmingly believed to result from local behaviour, with deforestation across urban and rural areas identified as the primary culprit. Climate change was perceived to be both caused by, and driving, this practice. As described by one resident who had observed an increase in deforestation:

> 'You expect that around March we shall start cultivating, it doesn't rain, or you plant and then it starts shining, you find that even in the villages people are buying food, yet those days people in the village were not buying food, they were getting food from their gardens. But you may dig a big portion, it starts shining, and the whole food [is destroyed]. So they are moving around cutting those big trees, now there will be an impact in future whereby those places become dangerous.' (Alfred, male, late 50s, Jinja)

Discussions of deforestation were accompanied by a sense of powerlessness in the face of the marginality and uncertainty of everyday urban conditions, forcing Jinja's residents into pursuing livelihood practices they knew to be unsustainable. As one younger man explained:

> 'You find that since the population has increased, most people have been forced to go into these things of trying to become hazardous. They have started chopping trees in order to sell them to get an income to sustain their families … Things have really changed in Uganda that

> even most of us have been forced to do things that we are not supposed to do. Things that are inhumane!' (Abdullah, male, late teens, Jinja)

Eguavoen (2013: 15) similarly found this issue to dominate his research on climate change perceptions in rural Ghana, with the community emphasizing how local tree felling is directly connected to changes in the weather but also how 'these practices were unavoidable as they secure their poor livelihood'.

From a local standpoint, climate change referred to weather changes occurring directly to Ugandans as a natural and/or spiritual consequence of human choices made there. Justine, in her mid-50s, described how changes in the weather simultaneously reflect both the will of God and human choices:

Justine: 'These days, you find that seasons keep on changing. Those days we used to plant our crops sometimes around this time, there would be not much sun like this one which is shining now. But now it's really shiny and very hot, especially at night, you can't even cover yourself. Even during the day, even if you're sitting in a place like this one, still you will feel when it's really hot. So I think there must be something wrong.'

Interviewer: 'And do you know why it is happening? Do you know what is causing it?'

Justine: 'I don't know. Some say that because of those factories, others say that because people have cut trees, others say because of those buvera [plastic bags] they throw anywhere. So I really don't know what it really is.'

Interviewer: 'What do you think personally?'

Justine: 'Me, ok on my side, me I think it's God, because God is the one who created this world. So he must be knowing what is taking place. Maybe that was his plan. And for me I think that the world is just growing old. It might not only be here in Jinja but even maybe in other countries. Some two days ago, I heard maybe where, in Uganda somewhere, it rained and then all houses fell down, then this lightening killed someone. So the rain was very bad … I don't know. It is really God who knows … Me, I just think, things are now getting worse. Everything is becoming worse.'

Interviewer: 'Worse in what way?'

Justine: 'Okay the way, we people are the people who are changing the world, I think. We are the ones who are becoming bad.'

These local explanatory narratives suggest a critical rereading of our survey findings. A sizeable majority of survey respondents in Jinja believed human actions are 'a major cause' of climate change, while in Nanjing and Sheffield residents tended to be more cautious and say that human actions 'definitely contribute'. Our interview and dialogue group data suggest that underlying this finding is a particular moral geography of climate change, rooted in a rapidly changing local urban setting and interwoven with the decisions people make (and are sometimes forced to make) in their everyday lives. Residents across Jinja, Nanjing and Sheffield might broadly agree that 'human actions' cause climate change, yet at the same time have very different perceptions of the extent to which this concerns the way they live.

"People shouldn't be ashamed that it's their problem"

Residents in Nanjing and Sheffield more typically exhibited an industrial blame stance on climate change, in particular one that focused on 'meta-emitters' (Cuomo, 2011) and the responsibilities of governments and industry. When asked whose responsibility it is to conserve the environment for future generations, they emphasized the national and global scale of climate change and therefore the coordination needed at this level:

> 'If the world still hesitates to take actions to protect the environment, the climate will get even worse … I think it's the government of countries around the world that should be responsible for that, because only they have the power to fix the problem.' (Zhenzhen, male, late 70s, Nanjing)

> 'People shouldn't be ashamed that it's their problem. Even if everybody in this country carries on in their own normal way, until big business stops pouring out its filth, we will have the issue about global warming and so on. Not so many years ago the Norwegian government were looking at why various types of fir type trees were dying off, and the reason was is that the gases from the Drax Power Station and Ferrybridge Power Station coal fire was drifting over there and it was killing their forest. We're not to blame for that … Until successive governments, and it has to be at that level, are prepared to challenge the multinationals – because that's what they are now – to stop pouring out the filth, whether that's China, America or this country or wherever, that will be one giant step forward.' (Daniel, male, early 70s, Sheffield)

These responses reflect a 'common sense' everyday morality, whereby people think that responsibility for harm is determined by the extent to which we directly contribute to it (Jamieson, 2010; Persson and Savulescu, 2012). Nanjing and Sheffield residents perceived that their personal impact on climate change is insignificant, and so held responsible national and global actors that can effect a bigger change for the better or worse (McKibben, 2012).

Nanjing residents particularly focused on how the Chinese government should balance its responsibility for environmental stewardship with the pursuit of economic growth and development, which was perceived to have come so far at significant environmental cost:

> 'Everyone has his own share of responsibility, but the larger part falls on the government: if the government wants to increase GDP and develop the economy, it will attach more importance to heavy industry which could affect the environment. The environment could benefit a bit from slow development, like there would not be so many cars … But some officials are so obsessed with creating economic benefits, realising outstanding political achievements as well as making more money that they would not care about the possible effects on the environment. I think that contributes most to the current problems. If the government could make some wise and compulsory policies, the public would follow.' (Bin, male, early 30s, Nanjing)

At a more modest scale, Sheffield residents similarly expressed support for regulatory approaches to enforce collective action on environmental issues. They offered examples of government initiatives that they perceived as largely successful, such as the provision of recycling infrastructure, the 5 pence plastic bag charge, and more recent microbead ban (Darier and Schule, 1999).

> 'The decision about plastic bags is an example of a fairly simple decision being taken, hasn't hurt anybody, hasn't impacted on anybody, it doesn't affect your free rights in any way … I think similar ideas in respect of renewable resources, which people won't notice particularly, if they were government-led could have a significant impact.' (Janice, female, early 50s, Sheffield)

The allocation of responsibility to regulatory authorities encompassed the idea that large-scale emitters in government and industry are not only more responsible, but more blameworthy for environmental problems (Lorenzoni et al, 2007; Cuomo, 2011; Rudiak-Gould, 2014).

These findings suggest that perceptions of responsibility are influenced by the scale at which climate change is imagined. In Jinja, narratives of self-blame were linked to a belief that environmental problems are locally rooted. In contrast, though Nanjing and Sheffield residents also conflated climate change with local weather and urban environmental change, they also envisaged its causes as global and complex. Thus, they reasoned that climate change is far removed from their sphere of influence and saw themselves as largely blameless for environmental problems (Jackson et al, 2009; Brace and Geoghegan, 2010):

> 'Keeping resources for the future? It is none of us ordinary people's business … Theoretically, it is still government, government's responsibility. What can the ordinary people keep? Is it useful for us to think about that?' (Huahua, male, early 50s, Nanjing)

> 'I'm talking about how cars pollute, but that's just one car. Down the road, there's a factory that has 50 years of pollution a day or something in comparison to the car. So it's a bit like, what level does my consumer choice actually affect?' (Karen, female, early 40s, Sheffield)

Feeling relatively powerless in the face of climate change (Gardiner, 2006; Lorenzoni et al, 2007; Marshall, 2014) was a prominent theme, as expressed by Sheffield resident Harriet:

> 'I don't believe what the media says about the environment. Obviously there's some truth to it somewhere along the line, but I'm a bit sceptical about what makes a difference … I know obviously dropping litter on the floor is going to make, that's a simple thing for me. But then taking it to that next level of how you could really make a difference, it's that that I don't really, definitely know, and also probably just push to the side of [my] mind … I know, obviously things that I know are true with the world in terms of the seasons and the changes in the weather and the ice melting and all that, I know that is true and that is happening. So from that point of view, when you start to think about that, and then it's quite a scary thought, to think that this is then probably going to speed up over time, how it will affect the world and how we pass it on to future generations.' (Harriet, female, early 30s, Sheffield)

While Harriet described herself as 'sceptical' of information about climate change, it became clear that she was not questioning whether climate change is happening, but rather, faced with overwhelming system complexity, was sceptical of claims about her ability to influence it (Maniates, 2004). Jamieson (2010: 438) claims that climate change is 'a rather deviant case of individual moral responsibility' because it is difficult to identify the 'causal nexus' between individual actions and the harm that, collectively, these actions contribute to. As Harriet's comment illustrates, people may feel indirectly implicated in the harm caused by climate change, while at the same time unsure of how they personally can make a difference. Hillier (2011) observes that the assumption of 'individual causal inefficacy' in relation to climate change is pervasive but also suggests, following

Parfit (1984: 77–8), that the belief that individual actions do not matter is a 'mistake in moral mathematics'. In the next section, we explore counter-arguments that seek to reconcile the global complexity of climate change with collective notions of responsibility.

"A tiny change brings a far reaching outcome": globalizing responsibility

The divergent moral framings of climate change employed here by citizens of more and less advanced industrialized nations are contrary to intuitions about global environmental justice. As such, they present an ethical challenge. While anthropologists such as Eguavoen (2013: 21) argue that climate change research should not focus on 'correct perceptions', the prevalence of self-blame among communities on the frontline of climate change provokes the somewhat paradoxical suggestion that they 'should be made aware of the actual causes of global climate change and not left believing that their management practices are responsible'. Rudiak-Gould (2014: 375) on the other hand suggests that indigenous narratives of shared culpability are important and, rather than displaying ignorance of anthropogenic climate change, they demonstrate 'that citizens can respond to climate change with something other than complacency and evasion of responsibility'. Jinja's residents emphasized their livelihood insecurity and "forced" choices in relation to environmentally destructive practices, but also felt that they could positively contribute to environmental stewardship through land management, avoiding waste and planting trees:

> 'We have to plant so many trees … and even to keep some areas like wetlands, to plant crops which are specific in that wetland not to build houses up as they have started, because that wetland also contributes a lot to the environment on

> rain maybe that's why the rain these days does not wet those wetlands, they also contribute but now we have destroyed them, we're building there houses which is wrong.' (Joy, female, early 50s, Jinja)

Through narratives that connected climate change with local behaviour, they believed that their actions have consequences. This raises questions about how researchers can work from perceptions of climate change culpability in contexts like Jinja, recognizing that cultural frames and local priorities may not be in accordance with Western models (Eguavoen, 2013). It also raises questions about how best to address perceptions of individual causal inefficacy among urban residents elsewhere, who envisage climate change through a global lens as a problem for *them, there and then* (Marshall, 2014): remote in space and time and/or too big a concern for 'ordinary people' (Persson and Savulescu, 2012).

This touches on a major ongoing debate about responsibility for climate change: the extent to which individual behaviour should be emphasized through universal blame rhetoric that 'shifts blame from State elites and powerful producer groups to more amorphous culprits like "human nature" or "all of us"' (Maniates, 2001: 43; see also Castree, 2008). McKibben (2012) for example argues that 'People perceive – correctly – that their individual actions will not make a decisive difference in the atmospheric concentration of CO_2'. Scholars have critiqued the responsibilization of individuals as agents of change as depoliticizing, particularly when their agency is envisaged primarily through their role as consumers (McEwan and Goodman, 2010; Shove, 2010; Middlemiss, 2014; Evans et al, 2017). It is naïve to suggest that all people need do is 'plant a tree' or 'buy a bike' to 'save the world' (Maniates, 2001); however, it is important to consider how to address the fact that industrial blame narratives can make people feel powerless to act on climate change.

While residents in Nanjing and Sheffield were reluctant to assume personal responsibility for protecting the environment, they also reflected critically on their eagerness to place the blame for climate change elsewhere. Harriet, who was sceptical of her personal impact on climate change, concluded this discussion by admitting "I'm probably wrong, because then actually if everybody thought that they could, and did something, then it's got to be better." Similarly, Nanjing resident Hui argued:

> 'Everyone should take an action, from the most mundane thing around him or her. I think through this way will make a big improvement, because we are only relying on the country and enterprises. I feel that we are just relying on the government.' (Hui, female, late 40s, Nanjing)

Our data suggests that residents in Nanjing and Sheffield felt some sense of responsibility for climate change, but also felt overwhelmed by it, believing that their personal impact pales in significance to the direction of travel set by governments and corporations. This sense of inconsequence cannot be met by proffering universal or local blame narratives of the kind that are so persuasive in Jinja. If people are to feel that they have some responsibility for climate change, they need a plausible reason to enact that responsibility in a way that is consistent with their diagnosis of its causes and key actors. For some of our interviewees, this reason came in the form of a global sense of place, connectivity and interdependency:

> 'The world is like a village, and China is a big country with a huge population. It's the butterfly effect, right? A tiny change brings a far reaching outcome. China has the biggest manufacturing industry. The way we produce goods and the marketing strategies will cause chain reactions globally.' (Zhang, female, late 20s, Nanjing)

> 'Well I now know through documentaries, through publications that some things are not good for the environment. Some habits that we carry on from previous generations we can't really carry on doing them because now we live in a global sort of world. Something that we produce here in Sheffield goes wherever. So yeah, one bit of the world affects the other. It's not like something is very local, something is compartmentalized. So – and I know that these things are not good and for the sake of me feeling well, I personally don't want to do these things.' (Claus, male, early 30s, Sheffield)

The global impact of 'a geographically and historically diffuse ecological harm such as climate change' (Cuomo, 2011: 697) challenges the assumption that people will meet seemingly distant threats with indifference (Massey, 2004; Barnett et al, 2005; Barnett et al, 2010; Popke, 2007; Persson and Savulescu, 2012). Smith (2000: 97) argues that 'understanding how "we", in the affluent parts of the world, impact on the lives and environments of distant others, can lead to an extension of a sense of responsibility'. The previous accounts of how urban residents' lives are connected with global networks of production and consumption suggest that such an extension is possible.

Some scholars are optimistic about the potential for global concerns like climate change to support the deterritorialization of 'ecological citizenship' (Spaargaren and Oosterveer, 2010: 1892), while others argue that it is counter to the human tendency to be spatially and temporally 'myopic' (Persson and Savulescu, 2012). Nanjing and Sheffield residents' concurrent belief in responsibility for climate change and the insignificance of their own actions illustrates how people can engage selectively with notions of connection and disconnection by 'invoking a politics of scale', emphasizing norms set by state and corporate elites to absolve themselves of blame (Jackson et al, 2009). Others have suggested that blame is not necessarily a helpful notion where a direct causal connection to harm is difficult to establish – as in, the harm caused by any one individual's contribution to climate change – but where there is nonetheless a shared responsibility for collective action (Gatens and Lloyd, 1999; Young, 2003; Barnett et al, 2005; Barnett et al, 2010; Jamieson, 2010). Young (2003: 42) terms this 'political responsibility' and argues that citizens have forward-looking responsibilities to address structural injustices that are connected with their lives. Similarly, Jamieson (2010: 433) argues that citizens have 'practical responsibility' for addressing climate change, observing: '[s]ince agents are causally efficacious both through individual and collective action, and through institutional roles, it is plausible to suppose that practical responsibility is plural and layered.' Young (2003:43–4) acknowledges that this conceptualization of responsibility can seem 'overwhelming' and 'unfair' – particularly when present injustices are at least in part a result of decisions made by past generations – but nonetheless holds that it is important to raise 'transnational and global questions about who should act for change'.

Despite Nanjing residents' emphasis on the importance of government policy relative to ordinary people's influence on climate change, they also reflected on the negative consequences for society if citizens are indifferent to environmental problems:

> 'Every detail in our life may not necessarily have a big impact on the climate but if the efforts of everyone are channeled together to be a form of positive energy, it will change things for the better … If everyone does not pay attention to it, there will be a negative impact on society which is for sure.' (Hua, male, early 50s, Nanjing)

Sheffield resident Terry expressed a belief in collective responsibility for climate change, arguing that it "doesn't matter" where manufacturing takes place insofar as it is unhelpful to blame China, when people in the UK and elsewhere benefit from the goods produced there:

> 'It doesn't matter where you manufacture these goods, does it? It doesn't matter where the impact is; it's planetary. People say oh it's over there. That's why we've shifted most of our manufacturing from this country … Most of our products come from China and all we've done is shifted that problem to China. They might be reaping the rewards of the wealth that they've created but they're not reaping the rewards of the climate impact that they've created as well. We all are, and that includes flooding, it includes all the things that are happening globally.' (Terry, male, early 50s, Sheffield)

This quotation exemplifies what Massey (2007) terms a 'politics of place beyond place'; a focus on global connectivity alongside local causality. In common with universal blame, the lens of political responsibility casts climate change as a shared problem for all of humanity. It makes it possible to talk of humanity's short-sightedness, while retaining a critical analysis of where power lies and how meta-emitters might be effectively held to account. In theory, this offers justification for action on climate change that addresses its 'overwhelming' politics of scale. However, there are challenges in translating this into practice, particularly in relation to responsibility over time, as we explore in the following sections on urban residents' accounts of intergenerational responsibility for climate change.

"That's the fear I have for the grandchildren": caring for the future

In public discourse on intergenerational responsibility, 'generation' is used interchangeably to refer to parent–child relations, to birth cohorts characterized by a shared generational identity, and to ideas about how people alive today ought to act in respect of the past and the future (Christophers, 2017; Little and Winch, 2017; White, 2017). Children are often cast as 'the future' (Phoenix et al, 2017) and parents and families are common tropes used to invoke social responsibility. While this family lens 'humanises the link between the past, present and future, and makes the complex sweep of time understandable' (Little and Winch, 2017: 137–8; see also White, 2017), the idea that this 'natural' chain of obligation between parents and children is scalable and equivalent to caring for the future is problematic (Persson and Savulescu, 2012). This framing is critiqued extensively in cross-cultural literature on intergenerational transfers and socioeconomic inequality (see Liu, 2014; Pöyliö and Kallio, 2017; Rowlingson et al, 2017). Others have noted its procreational and heteronormative overtones (Edelman, 2004; Vanderbeck, 2007; Little and Winch, 2017), how it privileges people 'like us', and reinforces an anthropocentric image of the future (Wu et al, 2007; Persson and Savulescu, 2012; White, 2017).

In recent years the rights of children, youth and future generations have become pivotal to the environmental movement's moral case for action on climate change, frequently invoked through emotional appeals to take action for 'our children' (Marshall, 2014; Little and Winch, 2017; White, 2017). These arguments have risen to prominence recently with the emergence of the international 'Youth Strike 4 Climate' movement. When asked which generation they think will be most affected by climate change, many of the people we interviewed in Jinja, Nanjing and Sheffield focused on its impact on younger generations:

> 'The children will be affected unless there is a change in weather, that the rains start coming as it used to be long ago.' (Faith, female, late 40s, Jinja)

> 'If the environment keeps deteriorating, the next generation will suffer the most.' (Daowei, male, late 40s, Nanjing)

> 'At some point, there's going to have to be a tipping point and thankfully we won't be around to see it, but that's the fear I have for the grandchildren.' (Gary, male, early 70s, Sheffield)

Societal narratives of time, Wallis (1970: 103) argues, 'motivate and interpret action', with the ways in which people imagine the future shaping how they act in the present. White (2017) proposes that the 'generational timescape' of climate change is simultaneously composed of 'the scale of the family' and 'the scale of humankind'. The former focuses on 'our' children and grandchildren and personal attachment to a not-too-distant future; the latter on future generations in the philosophical tradition of articulating moral obligations to distant past and future others. This generational timescape offers movement 'between the micro and the macro, the lived and the imagined, the familiar and the unfamiliar' (White, 2017:769). However, by casting future generations as kin groups, it shrinks the scope of ethical concern.

When 'our children' are used to persuade citizens of advanced industrialized nations to act, it is doubly problematic because most adults and children alive in these nations today are more likely to be shielded from the worst impacts of climate change. It detracts from the relationship between climate change and global inequality, reinforcing a 'limited and parochial altruism' (Persson and Savulescu, 2012). For instance, a survey by Leiserowitz et al (2017: 24) found that more Americans say 'providing a better life for our children and grandchildren' is, for them, the most important reason to reduce global warming than any other reason; ranked far above 'saving many plant and animal species from extinction' and 'saving many people around the world from poverty and starvation'.

When we asked broad questions such as 'Do people alive today have responsibility for future generations?' or 'What kind of responsibility do you think people have to future generations?' most residents responded by talking about their responsibilities towards family members:

> 'We have to work with the knowledge that we are working for our children in order for them not to suffer. Because if you don't work for your children, you are going to suffer and even the kids will suffer. But if you work for them and you make a firm foundation, our kids will never suffer.' (Brenda, female, early 20s, Jinja)

> 'Without a doubt. We do have responsibility. If you take it back to families, there's no doubt at all that families, parents are responsible for their children, to their grandchildren and so forth. We do have responsibilities; we can't just live for ourselves.' (Francis, male, late 80s, Sheffield)

The restricted spatiality and temporality of the family lens presents a significant challenge to envisaging responsibility for present and future climate change, as it is perfectly possible to care and provide for one's children in a way that is ecologically unsustainable. A very small number of residents reflected on this, arguing that their responsibilities to younger and future generations must extend beyond family members:

> 'We're not talking about my nephews or something. We're talking about children in other parts of the world who now haven't got enough food and access to water. As we know, it's a distribution thing, the food is, partly. But I think the climate change kicking in, it's going to be quite major

> really. So, those basics, I think future generations won't have.' (Helen, female, late 60s, Sheffield)

Many, however, highlighted how, in thinking about caring for the future, they focus on personal relationships because societal challenges like climate change are "too big" and render them feeling "helpless":

> 'Instead of advocating what we call a harmonious society, which is a too big thing for any individual to get hold of, we might as well work on personal relationships first. If every individual is able to do a good job in this respect, we will naturally have social harmony.' (Ying, female, late teens, Nanjing)

> 'You sometimes feel so helpless about it and I think one of the things about focusing on your own kids then is because you might be able to see something tangible from that.' (Sophia, female, late 60s, Sheffield)

This illustrates how the predominance of the family in framing idealized intergenerational relations of love, care and responsibility towards the future contributes to a particular generational timescape: a truncated view of time that privileges living generations in close proximity (Persson and Savulescu, 2012; Girvan, 2014; White, 2017), limiting engagement with global concerns and the 'long threat' of climate change (Dickinson, 2009).

"We want the day-to-day life": unimagining the future

In Jinja, residents spoke of focusing on intergenerational responsibilities at the scale of the family, but also how livelihood insecurity often meant not thinking about the future at all.

The future was difficult to imagine, they argued, given their preoccupation with the challenges of the present. "This gambling life" was a frequent refrain of young men in particular, articulating their sense of a lack of agency in the face of daily disempowering circumstances:

> 'Whether you think of it, but you don't have the guts to meet it. So what do you do? You just gamble and say that when I get 1,000 [Shillings] today, I'll eat that. The future will look after itself, that's where things are.' (Arnold, male, late 40s, Jinja)

Johnson-Hanks (2005) argues that across much of sub-Saharan Africa, political and economic crises have often served to sever firm links between intention and its fulfilment, making it risky to place one's hopes on a predictable midrange future. Mbembe (2002: 271) writes of a future horizon 'colonized by the immediate present and by prosaic short-term calculations' in which life becomes 'a game of chance, a lottery'. A focus on the everyday, rather than the future, in this way can have dire social and environmental consequences, for both resilience and individuals' perception of their capacity to fight against poverty (Appadurai, 2004; Pieterse, 2006; Henley, 2010; Duflo, 2013; Klein, 2014; Dalton et al, 2016).

New forms of 'short-sightedness' (Guyer, 2007: 409–10) in livelihood choices create the conditions of possibility for knowingly perpetrating environmentally damaging practices, with both immediate and long-term consequences for social-ecological relations. Samson, who ran a voluntary youth organization in one of the poorer neighbourhoods of Jinja, explained of the vulnerable youths with whom he worked:

> 'We teach them, but they have not gone to school and they have grown up and they are living in a small house

> with poor conditions. So he feels all he needs is to survive today, like think of what to eat and drink today. They feel they are failures already and they don't think about the future, they think they are already in their future … Yes they are already grownups, they feel like when I plant a tree it will grow in five years and I will be dead, they will not be there to see the future.' (Samson, male, mid-20s, Jinja)

With high morbidity and mortality levels, poor healthcare and increasing HIV prevalence, "living with lost hope" was generated when, as a Council Community Development Officer explained, "they think they will never become old. They think they are going to die when they are still young". An older woman reflected on the environmental impact of this in her neighbourhood:

> 'We want the day-to-day life. Cutting down the mango trees, the avocado trees, the jackfruit trees, this is done because of poverty. That's why people are cutting down trees. But they don't know what will happen after.' (Jackie, female, early 60s, Jinja)

These findings illustrate how, in the face of uncertainty, many of Jinja's residents felt compelled to prioritize personal and family subsistence rather than wider social and environmental sustainability. Within African studies, there is a growing recognition of uncertainty as potentially 'positive, fruitful, and productive', to the extent that it becomes a 'social resource' to be used 'to negotiate insecurity, conduct and create relationships, and act as a source for imagining the future with the hopes and fears this entails' (Cooper and Pratten, 2015: 2). Yet in Jinja, the entanglement of uncertainties generated by climate and socioeconomic changes contributed to an unimagining of the future.

"More of a now thing": imagined development trajectories

In Chapter Two, we discussed the relevance of respective nations' historical contributions to global greenhouse gas emissions to debates about environmental justice, contrasting the present day and cumulative carbon footprint of the UK compared to China. This is a key consideration in the moral geography of climate change and is enshrined in international policymaking. Gatens and Lloyd (1999) argue that notions of historical responsibility should encompass the benefits people enjoy today as a consequence of past development trajectories (see also Cuomo, 2011; Baatz, 2013):

> In understanding how our past continues in our present we understand also the demands of responsibility for the past we carry with us … We are responsible for the past not because of what we as individuals have done, but because of what we are. (Gatens and Lloyd, 1999: 81)

When we surveyed residents in Jinja, Nanjing and Sheffield, we asked about their views on a global polluter pays principle. As illustrated in Figures 4.3 and 4.4, the majority agreed both that 'Countries that produce more pollution like the UK and China owe a debt to poorer countries for contributing to climate change there', and especially that 'Countries that have historically contributed the most to climate change have a bigger responsibility to act today'.

However, in interviews Nanjing and Sheffield residents tended to be more ambivalent about the relevance of historical responsibility, citing excusable ignorance in defence of past generations who spearheaded industrial development (Caney, 2005; Baatz, 2013). When asked which generation(s) should take responsibility, they emphatically framed climate change as "more of a now thing":

Figure 4.3: Countries that produce more pollution, like the UK and China, owe a debt to poorer countries for contributing to climate change there

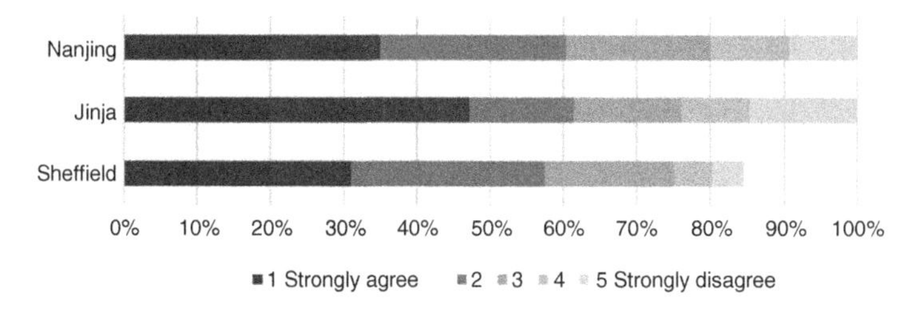

Figure 4.4: Countries that have historically contributed the most to climate change have a bigger responsibility to act today

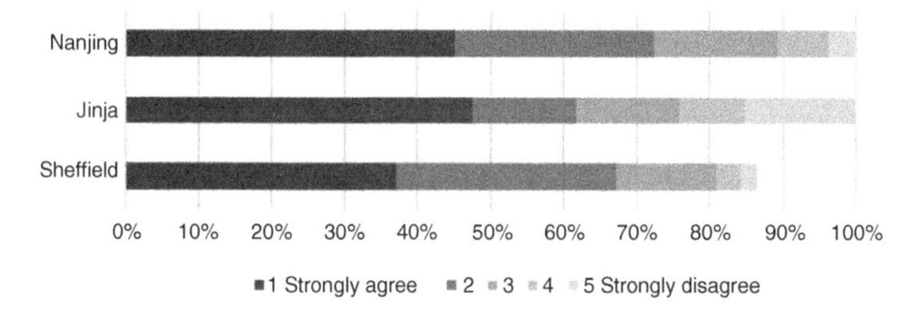

> 'It wasn't evident for the last generation, but bit by bit it has developed into the effects we are confronted with today.' (Daowei, male, late 40s, Nanjing)

> 'I don't think generations before really knew about it or considered it, so if you don't know about something you can't really prevent it. I think climate change is more of a now thing that we know that it's happening and we know that there's stuff that people do that cause it.' (Jill, female, early 40s, Sheffield)

This is a good illustration of how shared responsibility, rather than blameworthiness, is perhaps the more helpful focus for a moral geography of climate change. People who felt that they could not blame past generations for their decisions in respect of resource use nonetheless accepted that those decisions have had both benefits and negative consequences for people alive today. Furthermore, they accepted that people alive today are responsible for addressing the problems this has caused, with one important caveat that we explore next.

"If we want to develop the economy, we have to face pollution"

While industrial blame for climate change was prevalent in Nanjing and Sheffield, residents did not hold a wholly negative view of industrial development, often highlighting how it had benefited their national economies and living standards:

> 'In the past 30 or 40 years, our country has undoubtedly made great progress in economic development. But there is some associated loss, whether in foreign countries or in China … The ecological environment has changed a lot, in every aspect … Some may be very pessimistic, but I suggest we should be optimistic. We are all well aware of the Chinese Government's will to implement changes. It can do a lot of things in a short time.' (Fu, male, early 50s, Nanjing)

> 'I suppose you've got to look to the Industrial Revolution for the start of the issues, but if we hadn't had the Industrial Revolution we wouldn't have the security and the way of life that we've got now.' (Amy, female, early 50s, Sheffield)

In doing so, they contrasted the development trajectories of their respective nations and world regions in a surprisingly consistent narrative, suggesting that first comes economic development and then comes progress on environmental issues. In this respect, Sheffield and Nanjing residents alike envisaged their nations' development as interconnected, suggesting that China's recent growth could be likened to past industrial revolutions in Europe and elsewhere, and that Europeans now benefit from the displacement of manufacturing and its associated environmental costs to China:

> 'This [climate change] is because the society is making progress and the industry is developing. The changes are inevitable. But, in western countries, they also had such serious pollution 50 years ago, right? … China is developing in a way similar to other countries. If we want to develop the economy, we have to face pollution problems. After several years, the economy becomes good, and then we can invest to solve the problems.' (Chenmu, female, early 70s, Nanjing)

> 'China pollutes the atmosphere enough on its own, doesn't it, with its own manufacture? See we've got rid of all that. Sent all the manufacturing smoke and waste and dross over to the Far East to make our air better and but of course it's this – I don't know it's not ethical, isn't it to do that? But they've cleaned our country up.' (Sally, female, early 80s, Sheffield)

These narratives reflect an ecological modernization perspective, in which the economic pillar of development through growth and profit is initially prioritized, while the role of technology and innovation is emphasized in enabling transitions to more long-term sustainable development. While official Chinese discourses of *shēngtài wénmíng* (生态文明, 'ecological civilization') place greater emphasis on harmony with nature and quality of life than Western development discourses, they also present China as at the stage of rapid industrialization and urbanization and thus working towards sustainability (Liu, Chen et al, 2018).

Ecological modernization theory has been criticized by development scholars for neglecting power imbalances between nations (Lewis, 2000; Barry, 2003; Bailey and Wilson, 2009; Ewing, 2017). For instance, in Nanjing resident Lixin's account of how the Chinese government should pursue sustainable development, success is measured by displacing environmental problems to poorer nations, rather than actually solving them:

> 'I think they [the government] would like to have harmony between nature and the economy; but I think the economic development, after all, let's say, industry, this kind of thing, you need to work harder to develop industry to bring the economy up. Now is the era of being the big factory of the world … The only thing we can do is to follow other countries to develop your technology, to increase your technological level, and of course to enhance national strength, surely. And then, to be a bit nasty, we have to try to enslave other countries, based on our current ability, by transferring the pollution and other wasted stuff to them. And then, after your technological level is high enough, then you can achieve sustainable development inside your own country.' (Lixin, male, early 30s, Nanjing)

Drawing comparisons between China's industrialization and Sheffield's industrial past, Sheffield residents invoked the UK's historical contribution to climate change to suggest both that things will inevitably improve in China and that criticism of high emitting industrializing nations is "hypocritical". This somewhat provocative data highlights how thinking about global connectivity and historical responsibility for climate change can support a fatalistic response, valorizing technology and national capacity to displace these problems in the long term.

Chapter summary

Moral geographies of climate change are constructed at various scales, from a focus on local causality and self-blame in Jinja to emphasis of industrial blame, meta-emitters and global development trajectories in Nanjing and Sheffield. The scale at which climate change is imagined has implications for public engagement, for where it is perceived as remote or overwhelmingly complex, people are sceptical that they can make a difference and are thus less inclined to act. Following Young (2003), we have argued that focusing on political responsibility for climate change, emphasizing global connectivity, shared culpability and collective action, is one way of addressing this issue. We have also explored how, in thinking about caring for the future, appeals to act on climate change for 'our children' constrain imagined futures within relatively short-term generational timescapes, and how livelihood insecurities can contribute to unimagining the future. We have also considered urban residents' ambivalence with respect to historical responsibility for climate change, which in Nanjing and Sheffield is interwoven with beliefs about past ignorance, the inevitably of ecological modernization, and the power of their respective nations to displace ill-effects.

FIVE

Intergenerational Perspectives on Sustainable Consumption

> 'When my brother and I were little, although my father could probably have afforded to buy us whatever we wanted, he always got second-hand and did it up because that's the way he'd been brought up, to make do and mend. That's very much what it was; make do and mend. Well now everything is bought more easily and on credit maybe or not, but it's because it's instantly buyable in some way or other, I think they don't – it's not as appreciated as much and they just sling it out. Whereas we used to value things more and didn't expect to have as much.' (Marjorie, female, late 60s, Sheffield)

Introduction

This chapter considers what it means to consume sustainably in Jinja, Nanjing and Sheffield, in particular how residents' views on resource consumption are interwoven with anxieties about intergenerational value change (Inglehart, 2008). Sustainable consumption research has often neglected the Global South (Dermody et al, 2015; Ariztia et al, 2016; Liu, Valentine et al, 2018), or else it has emphasized the interdependencies of Northern consumers and Southern producers (Shanahan and Carlsson-Kanyama, 2005). There is, however, growing recognition of the globalization of consumer lifestyles and aspirations, and concern over the ecological implications of 'new consumer' trends in developing and transition countries (Myers and Kent, 2003). Our case study cities have very different histories and cultures of consumption, and we do not suggest equivalence in the impact or necessity of reducing consumption in each place. Rather, we are interested in how urban residents perceive local cultures of consumption and connect this with ideas about sustainable resource use. In this respect, we find common ground across Jinja, Nanjing and Sheffield in narratives of scarcity, frugality and waste, and in the characterization of unsustainable consumption as a generational problem.

"People who once lived a poor life can spend money on what they like"

At different paces and drawing on different embedded histories in each city, urban residents identified ways in which local cultures of consumption had changed in living memory. In Sheffield, these narratives centred on wartime scarcity and the post-war expansion of consumer choice:

> 'It was about 1954, 1955 I think we finished with coupons – food coupons – altogether. Then if you could get it, if you could afford it, it was available … It was ten years after the end of the war before things became reasonably plentiful again. I think this is why in the 1960s everything became so plentiful that everybody had excess. I wonder if it was our generation that went wrong. Because we'd grown up in our formative years being restricted. So did we, when we got the freedom, let our children – by that time my son was in his teens – did we let them do things that we couldn't do because they were not there? I do wonder if we started off this decline.' (Sally, female, early 80s, Sheffield)

In Nanjing residents spoke of the more recent transition to a market economy in the post-Reform era, with the youngest generation in our study the first to have been raised in this era:

> 'My consumption habits are different, for sure, from those of my sons and grandchildren. But in my opinion, my habits were developed and influenced by the poor economic conditions back in 1950s and 1960s. Besides, it was hard for me to raise my children at that time. I had to be frugal and thrifty. But China's economy has improved significantly, and my consumption habits have changed as I'm making more money. I'm also learning about the consumption ideas of the new generation … Chinese people are living a better life now, even people like me who once lived a poor life, can spend money on what they like.' (Zhenzhen, male, late 70s, Nanjing)

While both of these narratives focus on the expansion of consumer choice, they notably differ in their perception of this as a positive development, in Zhenzhen's case, or one associated with "excess" and "decline", in Sally's case. In Jinja, there was no such pivotal event that residents connected with changing cultures of consumption, but there was a cohesive narrative around the demands of urban modernity, an increasingly liberalized and capitalist local economy, and the rising significance of money for younger generations:

> 'The ancient people used to dig but for us in the modern generation we just wait for food from the market … We live a better life [because] we can decide for ourselves on the food we should eat … In the past they used to dig food for themselves and they could not decide for themselves the food that they are going to eat because they used to get food from the garden. Those ancient people did not want

> money because they had everything like food … so they did not want money as we do.' (Daudi, male, late teens, Jinja)

In Daudi's narrative, the marketization of consumption is cast in a positive light as expanding his freedom to choose. However, his reference to past generations "not want[ing] money as we do" is also suggestive of younger generations' experiences of urban livelihood insecurity generated by new economic pressures around generating cash-based income.

In each case, for better or worse, urban residents' narratives suggest that intergenerational value changes are taking place – or have already taken place – as a result of prevailing socioeconomic conditions. In the early 1970s, Inglehart (1971) hypothesized this in a Western context as a result of 'changing existential conditions – above all, the change from growing up with the feeling that survival is precarious, to growing up with the feeling that survival can be taken for granted' (Inglehart, 2008: 131). He characterized this as a shift from 'materialist' values that emphasize economic and political security in times of scarcity, towards 'post-materialist' values that emphasize autonomy and self-expression, which seemed to be becoming increasingly common among generations born after the Second World War. In his later work, Ingelhart finds longitudinal evidence of this hypothesized societal shift towards post-materialist values in Western countries and suggests that 'the logic of the underlying process remains relevant to much of the world' (2008: 137). Thus, he argues, in low income countries such as Uganda materialists tend to outnumber post-materialists, while those with rapidly growing economies such as China will approach a transition phase as younger generations' values are increasingly influenced by newfound prosperity, rising individualism and the emergence of a new middle class (Yan, 2009, 2010; Podoshen et al, 2011; Yu, 2014; Dermody et al, 2015). Similarly, Middlemiss (2014: 932) has argued that 'the advent of a consumer society (being able to create new 'selves' by buying things) [is] resulting in a profound change in our lived experience', with notions of sustainable consumption related to the idea that people are becoming increasingly individualized and turning to the market to express themselves.

Generational blame for unsustainable consumption emerged, as we discuss in the next section, as a prominent theme in our interviews with residents across all three cities. However, in our survey it was Jinja's residents who were most concerned about a rise in materialist values among the younger generation. As Figure 5.1 illustrates, almost three quarters (74 per cent) of survey respondents in Jinja agreed or strongly agreed that 'Our environment suffers because the younger generation is more materialistic than previous generations', compared with less than half (47 per cent) of respondents in Sheffield and fewer (42 per cent) in Nanjing. This illustrates the importance of rising materialism as a key narrative frame in Jinja. In Sheffield, although there was weaker agreement with this statement than in Jinja, more than twice as many local residents agreed than those who disagreed. Many Sheffielders shared Jinja residents' anxieties about the younger generation's "great obsession with money". However, they differentiated between past generations' pursuit of money for economic security and a present situation that one Sheffield teenager summed up as "the richest are getting richer and people are obsessed with material goods rather than just a good quality of life". In other words, they saw materialist values and conspicuous consumption as persisting and coexisting, somewhat uncomfortably, alongside post-materialist concerns such as happiness and self-fulfilment. Opinion on young consumers was most divided in Nanjing, perhaps reflecting residents' more positive view of materialist values in light of China's recent economic growth.

Inglehart's theory of intergenerational value change helps to explain why consumption is cast as a 'generational problem' (Vanderbeck and Worth, 2014). Scarcity and choice emerged as prominent themes in urban residents' narratives of how local

Figure 5.1: Our environment suffers because the younger generation is more materialistic than previous generations

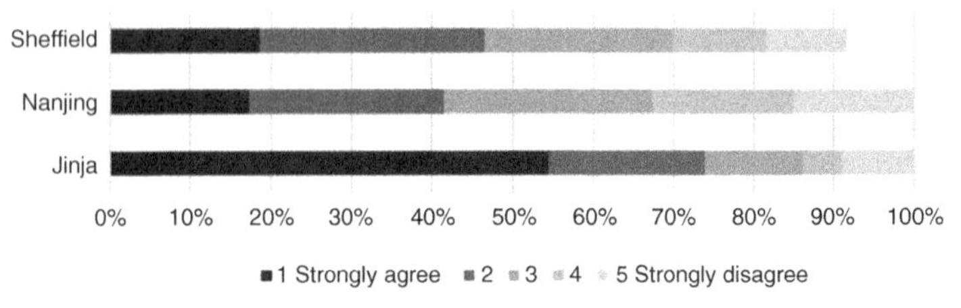

Note: An additional 'Don't know' response category was available in Sheffield only.

cultures of consumption are changing. Residents in Nanjing and Sheffield described the overwhelming range of consumer choices available and the impact this has on younger generations:

> 'People today? It is rather hard for them, perhaps. This is because there are just too many choices for them today. There are hundreds and thousands of clothes for you to choose, when you want to buy.' (Yu, male, early 20s, Nanjing)

> 'They [older generations] grew up in a time where they didn't have much, so they're probably more sparing. Whereas people today are generally … quite well off compared to how people used to be. So we take a lot of things for granted, I think, because we've just got it so easy. If we need something we just go out and buy it, within reason.' (Sonia, female, early 20s, Sheffield)

In contrast, residents in Jinja cast younger generations' pursuit of money more in the context of enduring urban poverty, resource scarcity and the need for riskier livelihood strategies amidst a rapidly expanding urban population (McQuaid et al, 2018a):

> 'The first generation they were few and there was not much need because they were just using these small, small

> branches, not cutting the whole tree down so they were preserving the environment for the future, but now they have to use whatever is there … You have nothing to do, you have to cut that tree because you will get money for us, you cut that tree because you want to cultivate within that area, so this generation know, but they have nothing to do.' (Benjamin, male, late 50s, Jinja)

In Benjamin's account, the unsustainable rate of tree felling for charcoal arises from lack of choice and satisfies a material necessity, whereas in Sonia's and Yu's narratives unsustainable consumer lifestyles are the result of "too many choices". Yet these narratives share a tendency to cast younger generations as the chief culprits of unsustainable consumption, expressing shared anxieties about an intergenerational shift in human-environment relations. We explore the key features of this generational blame narrative in the next section, and how this is juxtaposed with idealized accounts of past generations' responsible resource use.

"This generation doesn't have the patience to leave the tree standing"

The previous chapter on moral geographies of climate change explored various blame narratives including generational blame (Rudiak-Gould, 2014; Hulme, 2017), urban residents' ambivalence with respect to historical responsibility for climate change, and their emphasis on it being "more of a now thing":

> 'I think it's this generation now that where the facts are there for the first time, there's evidence there and it's our responsibility more than anyone that's ever been to act on that.' (Joe, male, late teens, Sheffield)

Related to this emphasis of the present and the idea that past generations cannot be blamed for climate change was a tendency to idealize the sustainable lifestyles of past generations. When we asked which generation is most likely to consider the environment, people in Jinja and in Sheffield especially made unfavourable comparisons between their generations' resource use, and that of their grandparents' resource conservation:

> 'In my family the generation that has really considered the environment is that of my Jjaja's [grandmother] because for them they always struggle to see that something they found there is still there. Take an example of the Mabira forest; it was going to be destroyed so that [the] Madhvani [company] might plant sugar cane but it is the old people who woke up and said, "No you are not destroying this forest, we found it there and we want other generations to come and find it there and also study about it." … In this generation people are not patient so they always want short-term things and there is too much corruption in this generation.' (Amaya, female, late 20s, Jinja)

> 'After the war had ended, because of the rations and things like that, everyone was high on watching out what they were doing … Then as we've gone away from the war we've been more well off, and because of that people have been less concerned about it. I think it's a trend. If we are in difficulties – if a family is in difficulty, they'll watch what they do like a hawk, so they recycle, they do everything correctly. But once the difficulties end, they start buying things, the luxury items, again. Because my Nan-Nan [grandmother] grew up in a more difficult time, she watches out what she's doing more.' (Tyler, male, late teens, Sheffield)

A key moral frame in their narratives of climate change was a discursive construct of past generations living frugally and in harmony with the environment. These idealized accounts were held up in protest against current economic and climatic conditions and the perceived profligacy of younger generations, to 'paint a nostalgic picture of the past, and an apocalyptic vision of the future' (Middlemiss, 2014: 939) in which, it was feared, people will become increasingly selfish and individualized. In Nanjing, people talked about social change and the expansion of consumer choice in more positive terms, but nonetheless associated this with the acceleration of environmental problems:

> 'People of this generation … we live in a fast-paced development era. Some hidden problems are left by people of the past generation, but we let these problems keep growing.' (Sijia, female, late teens, Nanjing)

Today's economically active generations were thus characterized as careless with resources, and more complicit in climate (and environmental) change than their forebears, because they are careless in spite of improved knowledge about environmental management.

In Jinja, rising inequality was perceived to be entrenching short-sightedness whereby the pursuit of material goods is being prioritized over environmental conservation, and, as discussed in the previous chapter, discouraging future-orientated thinking:

> 'People today, they love a lot of money. This generation doesn't have the patience to leave the tree standing, but our parents, for them they were respecting their [grandparents'] things, like they knew that if you plant a tree, you have like a fruit tree you are supposed to have a fruit but not cutting it down … In my village, the problem came about when people found out that you can get money for charcoal … The generation now don't prepare for future generations, they look at how so-and-so built a storied house, they want it, but they don't prepare for their own children.' (Betty, female, late 20s, Jinja)

Residents emphasized changes including lower levels of competition for land and resources, which enabled past generations to achieve subsistence affluence, planning for, acquiring, improving, and transferring environmental assets (particularly land) to the next generations of their family and clan. Discussing this theme in a generational dialogue group, a middle-aged man explained how "the environment was part of us, our culture, but now our culture of environmental conservation is being eroded". In their view, a community perceived to have once been able to live in harmony with the environment – and thus enjoying predictable and sustainable rain patterns – has been replaced by a new generation corrupted by what was alternately (or sometimes simultaneously) represented as self-serving individualism or desperation to survive in a new context of the urban cash-based economy, poverty and reduced opportunities:

> 'My mother's generation is the one which has been thinking about the environment and conserving it, but our generation, we are the people who are spoiling everything … I think it is poverty.' (Rose, female, early 30s, Jinja)

Residents argued that new livelihood practices for generating money were both causing climate change and violating past generations' standards of behaviour. Here, climate change narratives intersect with familiar narratives of moral decay in the African city (Porter et al, 2010; Sommers, 2010), with extreme weather events such as droughts and storms perceived as proof of a rupture in the natural order.

In Sheffield, narratives of unsustainable consumption likewise linked environmental degradation with moral decline. This is exemplified in Mary's quotation at the beginning of this chapter about how her generation "used to value things more", and Sally's account of how her generation indulged their children in the post-war years of prosperity. Similarly, Sheffield resident

Lester, who described himself as "from a reasonably poor family [who] used to make do and mend", reflected on why, in his view, this habit has not been passed on to his grandchildren:

> 'I think it's consumerism gone mad. It's to do with big business just pushing people into buying stuff that they don't really need. And it affects most people; they'll end up with – they'll go out and buy a car and they don't really need a car because they've got one that's perfectly good and instead of repairing it they'll go and get a new one. They can't afford it. It's not good for society. It belittles it a bit.' (Lester, male, late 60s, Sheffield)

Rising prosperity hand-in-hand with inequality was perceived to have resulted in corrosive greed that is contributing to human misery (Jackson, 2005; Jensen, 2013; Forkert, 2014). Every generation since the Baby Boomers was characterized as more consumer-driven and wasteful than older people who grew up in times of scarcity (Robins, 1999; Carr et al, 2012). These findings reflect a prominent societal narrative in Western contexts, which Evans (2011a: 42) describes as the 'throwaway society thesis' that contrasts the 'excessive, wanton nature of contemporary consumerism as compared to an earlier time in which our thrifty forebears were (imagined to be) far less profligate'. This narrative has been critiqued as a pervasive and powerful 'myth of consumerism' that reinforces conservative social norms, and scholars have questioned the extent to which it withstands empirical scrutiny (Gregson et al, 2007; Evans, 2011a; Forkert, 2014). Nonetheless it has an undeniable influence over public opinion.

In Nanjing, residents were less critical of younger generations' consumption habits and more positive about post-Reform changes in consumer culture. Some, however, were anxious about the impact of newfound prosperity on younger generations. Echoing Sheffield's post-war narrative of scarcity followed by increasing indulgence, they suggested that consumption habits will change for the worse in the future as more people grow up comparatively unrestricted. Yangyang, a parent in her early thirties, discussed how she finds it hard to teach her child the importance of conservation when resources seem plentiful:

> 'It's assumed that for the next generation, materials will be abundant, and there will be no lack of supply, as long as you have money. They probably will not be conservative consumers ... The next generation will not consider conservation when there are abundant materials available for them. I once noticed that my child forgot to turn off the tap after washing and tooth-brushing even when I told him to. He has no idea how rare resources are. He just does not care.' (Yangyang, female, early 30s, Nanjing)

Feng felt that younger people are becoming more careless in their spending, for example more likely to get into debt, and speculated that his own children will follow suit:

> 'I think they will be less rational on consumption. I often buy whatever they [my children] want. Sometimes I think in the future they may turn into impulsive shoppers. If we would like to influence them, when shopping, we should select only the necessities for them.' (Feng, male, late 20s, Nanjing)

In China, anxieties about intergenerational value change in the context of consumption can be related to the 'Little Emperors' phenomenon and the child-centred family structure that emerged following the launch of the one child policy in 1979 (now relaxed to two children in some circumstances). Little Emperors 'receive much more than Chinese children of previous generations could ever have dreamed of' (Shao and Herbig, 1994: 16) and, it has been argued, 'often grow up spoiled, demanding what they want and spending what they have' (Wang, 2009). Especially in urban contexts where children's education is prioritized, they can expect to gain professional employment and earn good salaries as they enter adulthood, joining the expanding ranks of the young, emerging middle classes who place increasing emphasis on instant gratification and luxury goods.

"The grannies are still planting": idealized sustainable lifestyles

In contrast to these narratives of generational blame focused on younger generations, residents' accounts of what it means to consume sustainably often emphasized traditional virtues and idealized the sustainable lifestyles of past generations. City residents perceived younger generations to be less knowledgeable about living in harmony with the environment and lacking subsistence skills, evoking a halcyon past. For instance, Deirdre, who grew up in a farming family and came to live in Sheffield in later life, argued:

> 'I think once a farmer, you're always a farmer. You've got that instinct to use it up and feed your babies well … and you're always planting things. Always making sure that the garden, that's a very important part of it, was nurturing your garden because that was going to be your food.' (Deirdre, female, late 80s, Sheffield)

Similarly, other older residents discussed sustainable consumption in terms of subsistence lifestyles and agricultural knowledge:

> 'The grannies they are the ones who were caring, they still care because they are still planting, they are still planting. Those people were caring about the environment.' (Rebecca, female, early 60s, Jinja)

> 'Food is necessary for everyone. However, few people know how to farm nowadays … If we don't pass down our knowledge about agriculture to the next generation, they'll lose the ability to farm. Everyone needs food. Who's going to plant grain for us all?' (Hanyang, male, early 50s, Nanjing)

Yet it was only in Jinja that peri-urban agricultural practices persisted to a significant degree in the everyday lives of local residents, with legitimate concerns that younger generations are not returning to the villages to carry out full-time agriculture and risking their future food security (Eguavoen, 2013). In Sheffield, allotments and food growing featured in the wartime narratives of older residents and among a very small number of typically older people who owned or leased land, but as in Nanjing agriculture was a marginal activity. Notions of subsistence-based sustainable livelihoods therefore evoked an imagined past premised on the construction of a 'rural idyll … as a counterpoint to urban/industrial life and work' (Rigg and Ritchie, 2002: 361), often drawing on the experiences of rural to urban migrants, rather than reflecting actual differences in contemporary generational practices within these cities.

Jinja's residents' cultural model of climate change blame included, related to their emphasis on local causality, a belief that changing social moralities and intergenerational relations are in part responsible for the changing weather. Across the African continent urbanization and accompanying economic shifts are slowly eroding gerontocratic intergenerational 'contracts' in which established rules on the transfer of resources and responsibilities are breaking down (Collard, 2000; Durham, 2000; Burgess and Burton, 2010; Frederiksen and Munive, 2010; Banks, 2015) or shifting in form (McQuaid et al, 2018b). At the same time, most older residents of Jinja have experience of a more rural, subsistence-based way of life. In contrast, many members of the younger generation have only experienced urban living. In this context, Jinja's residents – of all ages – contrasted contemporary consumption practices and livelihood strategies with more sustainable livelihoods in the past, which used to provide a foundation for future generations. The demands of urban modernity (including the increasing significance of money) and decline in livelihood opportunities were perceived to have disrupted intergenerational solidarities that once acted as key social protection mechanisms. Residents described how, in their view, people were becoming *"harsh"*, *"no longer helping"* and increasingly focused on their own needs and wants:

> 'They just construct, they just work for the day-to-day money. People are not planning for the future or their family ... People are just looking for money and caring about their stomachs ... We need to teach the young people to think about the whole country, then we can't just think about ourselves.' (Jackson, male, mid-40s, Jinja)

The older generation's ecological knowledge, steadfast traditionalism and communal values were juxtaposed with contemporary consumption practices, which were perceived as short-sighted and self-serving, posing a risk to both social cohesion and the environment. This highlights 'the importance of internal moral discourses about the production of disorder and stagnation' (Eguavoen, 2013: 20). For Jinja's residents, sustainable resource use meant "sensitising" younger generations to the importance of environmental conservation, passing on agricultural knowledge, planting trees, preserving land instead of selling it, and reinstating a culture of sustainable human-environment relations.

"Make do and mend" and "qínjiăn jiéyuē"

Although their particular consumption concerns and practices were very different, residents in Nanjing similarly defined sustainable consumption in terms of their responsibility to save resources for future generations:

> 'For people in our generation, we can't say everything belongs to us or to this generation. Whether knowledge, wealth, environment, nature, and animals, we need to have the concept of passing it on. Isn't that right? We need to do better, leave things with certain values to the next generation, and try our best to make this society and nature more prosperous. It's not an option to take and use up everything.' (Zhang, female, late 20s, Nanjing)

Their narratives were less focused on preserving land, however, and more on acting rationally through *qínjiǎn jiéyuē* (勤俭节约), a traditional Chinese cultural value that translates as being diligent and thrifty or frugal (Wang, 2009). Being *qínjiǎn jiéyuē* involves spending money according to one's income and needs while avoiding overspending, wasteful consumption and impulsive purchases. Zhenzhen, previously quoted discussing how his consumption habits have changed as socioeconomic conditions in China have improved, still emphasized the importance of being thrifty when making purchase decisions:

> 'Thrift is a virtue. I think one should show this quality at all times. But thrift isn't the same as parsimony. It doesn't mean you shouldn't buy something because it costs money. Thrift is buying what you need while not squandering money.' (Zhenzhen, male, late 70s, Nanjing)

Qínjiǎn jiéyuē is strongly interlinked with Chinese moral teaching, such as the Confucius saying *wúyù zé ang* (无欲则刚, 'without desires, one will become resolute') from The Analects (Slingerland, 2003), and Maoist ideologies that contend that consumerism is a false way to happiness, valorizing restraint as a corrective check on materialist values (Sun et al, 2014). *Qínjiǎn jiéyuē* is also aligned with the practice of saving, as discussed by Yang in her comparison of Chinese and European consumption practices:

> 'I think Chinese people generally live a thrifty life, which is a traditional virtue in China [and which] has some positive influences for sure. During the European debt crisis for example, many Europeans have no financial saving awareness, and I think Chinese people are doing better in this respect.' (Yang, female, mid-20s, Nanjing)

In a Western context, scholars have questioned whether thrift should be connected with sustainable consumption, as it is more clearly linked with economizing and saving money, freeing up personal resources for further consumption (Gregson et al, 2007; Pepper et al, 2009; Evans, 2011b). While the majority of Nanjing residents' accounts of *qínjiǎn jiéyuē* consumption did emphasize financial considerations such as cost-effectiveness and saving money for oneself and one's family, they also expressed a deeply ingrained sense that it is a moral obligation for the good of society. The specific cultural characteristics of *qínjiǎn jiéyuē* are thus more suggestive of a social practice view of consumption than one rooted solely in individual rational choice (Warde, 2005; Shove et al, 2012; Liu, Valentine, et al, 2018).

There are some parallels between the Chinese discourse of *qínjiǎn jiéyuē* and the valorization of thrift in the UK context through the appropriation of wartime slogans such as 'make do and mend' that encourage living with less. This was a frequent refrain in older Sheffield residents' accounts of sustainable consumption, which emphasized how their values were shaped by conditions of scarcity such as wartime rationing and/or poverty. This was perceived to have resulted in their being careful with limited personal resources and the capacity to save, reuse and manage on a budget:

> 'I've always wanted a little bit of a margin, as we used to call it, between what we have and something in the bank for a rainy day. Yes, I've always wanted to have something put by ... I think there was such a lot of make do and mend after the war – it makes me sound old but indeed I am old, so yes. So one was brought up with that and that's hard to release, I think. But the next generation, that wasn't necessary, so they don't have that background.' (Jeanette, female, late 70s, Sheffield)

We also found evidence of thrift appealing to middle-aged and younger people, who placed greater emphasis on thrift as a lifestyle choice motivated by aesthetic and environmental considerations as well as saving money:

> 'The generation below are going to have to almost revert to how my parents and grandparents thought. There is that thing of make do and mend and recycling or preloved stuff coming back in, isn't there? I think that's almost going back to the [19]50s.' (Faith, female, early 50s, Sheffield)

> 'I'm happy to have second hand clothes, second hand – our house is second hand furniture, clothes. I am a make do and mend sort of person. I'm – charity shops and vintage shops – quite happy to shop like that. I like my old lady furniture anyway, so when we actually move it will be kitted out like 1950s [...] I think as long as you have got people like us willing to accept second hand stuff, then I think that's a good – it's less impact on the environment.' (Gemma, female, early 30s, Sheffield)

In the recession that followed the 2008 financial crisis scholars noted the emergence of a 'new thrift' trend in the UK, suggesting a cross-generational tendency to focus on economic security in times of austerity (Evans, 2011b; Jensen, 2013; Forkert, 2014). Inglehart (2008) argues that socioeconomic shocks can have short-term 'period effects', but that generational cohort values are relatively stable overall. Thus, younger Sheffield residents tended to emphasize saving money less than their elders. They re-appropriated thrift as a post-materialist value concerned with identity expression and caring for the environment, a strategy for 'managing greed' and questioning the link between conspicuous consumption and wellbeing (Jackson, 2005). As Roxy, in her early 30s, argued: "I like to live – it's not – well, it's frugally to a point, but I think about using what I need, not what I think I should have." This suggests that, in the Sheffield context, sustainable consumption is connected with both a materialist emphasis on being careful with and freeing up personal economic resources, and a post-materialist emphasis beyond saving money 'related to deeper notions of being frugal and treating resources with care' (Barr et al, 2011: 3019; see also Evans, 2011b).

"All I see is landfill sometimes": the totemic role of waste

This chapter has so far focused on the local characteristics of perceived intergenerational value change in consumption practices across Jinja, Nanjing and Sheffield. However, one key concern that united residents of all ages across the three cities was the role of waste in making unsustainable consumption visible. Many residents shared anxieties about plastics, litter and the everyday management of waste, from the dumping of *buvera* (plastic bags) in Jinja to the amount of packaging on supermarket groceries in Sheffield:

> 'Then the other pollution, environmental pollution I can see is these, eh! What do you call them? These buvera … the polythene bag. Being the estate having a lot of people, these people are a conduit, because whenever they go to the market they are packed into those things. Now you find that they are everywhere and at times most of them find their way into the waters and the land around, or the dustbin around here is full of those polythenes, which I feel are not good for the soil here.' (Musa, male, early 40s, Jinja)

> 'I barely see paper bags in China. We all know that plastic bags are made of PE [polyethylene] or PPE [poly(p-phenylene oxide)] and it is not easy to biodegrade even after hundreds of years … Undoubtedly, it is harmful for the environment. However, plastic bags are widely used in China. Many people use plastic bags, including me.' (Jin, male, late 20s, Nanjing)

> 'I walk into a shop and all I see is landfill sometimes. There's just so many pointless things and the amount of packaging that they have around products is just completely unnecessary. Everything's just coated in plastic and there's no need for it.' (Lizzie, female, early 20s, Sheffield)

Other researchers have similarly noted the prominence of waste, especially objects such as plastic shopping bags and packaging, as symbols of consumerism (Shanahan and Carlsson-Kanyama, 2005), 'destructive excess' (Crang, 2012), and the throwaway society. They argue that such objects occupy a 'totemic role in the debate about sustainable consumption' (Peattie and Collins, 2009: 110) that is disproportionate to their actual impact. Gregson et al (2007) suggest that this is because such waste is especially visible to urban residents through the management of household resources and the everyday appearance of their neighbourhoods, and thus provides a tangible way for them to relate to the idea of sustainable consumption.

The extent to which waste and plastics dominated discussions of sustainable consumption was also apparent in theatre workshops across the three cities. In Nanjing, residents developed a short play called 'Supershop' that used plastic bags and balloons to represent consumer goods that a magical shop could produce for free on demand, symbolizing the transient nature of consumer desire. In Sheffield, residents were asked to bring unsustainable objects as talking points in both interviews and theatre workshops and tended to choose things like plastic bags, plastic bottles and mobile phones, observing that "People chuck away their phones, their mobile phones every couple of years". In this respect, they shared concerns with Nanjing residents about the proliferation of consumer electronics with a short lifespan:

> 'What I regard as a waste is like frequently replacing cellphones, which I don't think is necessary. Many people today prefer buying high end products, like iPhones ... These things may satisfy your needs in some way, but they are not what you must have. I think buying anything which is not necessary, or which only charms one's personal vanity, is a waste.' (Ying, female, late teens, Nanjing)

In Jinja, a facilitated process of intergenerational interviewing and creative environmental knowledge sharing (discussed as a case study in Chapter Six) identified residents' common concerns about the lack of waste management infrastructure, as well as deforestation and household energy security related to the rising price of charcoal. These issues formed the basis of a creative action research process in which residents learned how to reuse household organic waste and charcoal dust from cooking to make briquettes, turning "waste into wealth" as an alternative source of biomass fuel. A Green Briquette cooperative was formed as a result of this process and residents organized a number of waste action interventions in neighbourhoods around Jinja, using intergenerational community-based research and applied theatre to engage others in thinking about the identified environmental issues and the potential for reusing household waste.

In Jinja, the council authorities are supposed to collect waste from commercial sites but disposing of household waste is the responsibility of local residents, and much of this waste is buried or burned. Residents were thus able to intervene directly, albeit on a modest scale, in local waste management through the Green Briquette cooperative. In Sheffield and Nanjing, in contrast, the main ways that residents engaged in waste management was through recycling: 86 per cent of survey respondents in Sheffield and 90 per cent in Nanjing said that they recycle, and many interviewees were eager to explain what and how they recycle at home through municipal waste collection and other means:

> 'My husband goes mad because I've got a box here, I've got this box here, I've got my rubbish bin. At the side I've got my recycling bin. Then I recycle plastic tops for Cancer Research. Then somewhere else, clothes go to Salvation Army, Archer Project.' (June, female, late 60s, Sheffield)

In such narratives, recycling was framed as a waste avoidance strategy connected with thrift and social responsibility as well as environmental concern. For example, June's account mentions health and homelessness charities that will benefit from things she might otherwise have thrown away. Her account is also interesting in highlighting how recycling has become more of a priority for her in later life. When asked if she has always recycled, she replied:

> 'I think since it's been – you know like the Council, I mean no, not when we were younger because it wasn't done,

> it wasn't done. It's only since they started saying you can save this or you can save the environment by doing this or by doing that.'

Social practice theories of sustainable consumption emphasize the role of routines, habits, social norms and infrastructures, suggesting that these influence everyday consumption more than the values a person holds (Dolan, 2002; Shove, 2003, 2010; Warde, 2005). Recycling offers a good example of this theory in action, illustrating how certain practices can become embedded across generations. In recent years the UK's waste policy landscape has 'demonised landfill and promoted recycling' (Gregson and Crang, 2010: 1027). Changing waste disposal infrastructure and public information campaigns have established a new social norm and drawn householders' attention to what they are throwing away, resulting in a threefold increase in English household recycling rates since 2000 (DEFRA, 2015). In this context, our data suggests that people of all generations are mindful of waste and what they ought to recycle, regardless of their identification with explicitly pro-environmental values.

In Nanjing, residents' ideas of *qínjiǎn jiéyuē* consumption encompassed reducing waste, with many citing the saying *làngfèi kěchǐ* (浪费可耻, 'it is shameful to waste') when asked what it means to consume sustainably. The link between *qínjiǎn jiéyuē* and reducing waste is evident in public discourses of sustainable consumption in China. In 2001, the Chinese Communist Party's Sixth Plenary Session of the Fifteenth Central Committee announced its promotion of the construction of a resource conserving society, outlining a national vision of transition towards a circular economy, and encouraging moderation in personal consumption predicated on saving energy, food and resources, and reducing household waste. In addition to plastics and electronics, many Nanjing residents discussed strategies to avoid food waste, such as ordering less and keeping leftovers, in reference to a national *guāng pán xíngdòng* (光盘行动, 'clear your plate') campaign promoted by both the Government and national NGOs. However, some residents emphasized how other concerns take precedent over avoiding waste, such as the importance of *miànzǐ* (面子, 'saving face') culture in the workplace and when they are with their peers:

> 'If you order a lot of food when you dine out or go out with friends, and you can't finish it, but you want to save face, so you won't wrap it up and take it away … I think people generally overconsume a lot. Since I have money, I don't really care.' (Junyi, male, mid-20s, Nanjing)

Thus, while saving and not wasting is a prominent national discourse of sustainable consumption in China, this is not always reflected in everyday life. *Miànzǐ* is one example of an alternative driver of materialistic aspirations and conspicuous consumption among young urbanites, with the appearance of being able to consume frivolously conferring social status (Sun et al, 2014) and resulting in waste.

Chapter summary

It is challenging to talk about sustainable consumption with parity across national and local urban contexts where there are stark inequalities in terms of global wealth distribution and market access. However, it is helpful to move beyond the dichotomy of Northern consumers and Southern producers, to understand the meanings and practices of (un)sustainable consumption in emerging and transition economies. Our research has focused on the cultural construction of sustainable consumption in the everyday lives of urban residents, in particular how it is commonly framed through moral panics that express anxieties about (inter)generational value change. Generational blame for unsustainable consumption was a common thread across diverse cultural narratives of living with scarcity and excess, which included concerns about rural to urban livelihood transition, rising materialism, the disruption of social solidarities, the devaluing of subsistence skills, the indulgence of children, and the acceleration of 'consumerism gone mad'. In contrast, older generations were perceived to have led more sustainable and virtuous lives. We do not suggest that these narratives are representative of the actual consumption practices of older and younger generations, rather that they offer an insight into how perceptions of sustainable consumption are interwoven with multi-scalar experiences of socioeconomic transition. We have also highlighted the prominence of waste as a key sustainability concern among urban residents, and how it functions as a potent symbol of consumerism that amplifies anxieties about irresponsible resource use. In the final chapter, we consider how such anxieties might be addressed through intergenerational community-based research and arts practice to encourage critical reflection on sustainability.

SIX

Imagining Alternative Futures

> The young generation must first learn/ how to love and respect where they live/ for, that is where the environment starts … Our children, please control the land!/ This is our duty, to tell this generation/ to use properly what we have,/ how to use the small land we have. (Extract from 'We Are the Foundations', a verbatim poem composed from older people's words in Jinja)

> Our beautiful home is disappearing, the rivers are dirty, the air is being polluted, the green hills are withered … Us humans, we should treat nature with kindness, work hard and protect our home to awake the dying land. (Extract from *Homeland*, a shadow puppet play written and performed by older people in Nanjing)

> We all played our part in getting to this throwaway, just a click away world/ Buy one get one free – free from sweatshops, extinctions, pollution and climate change? … Looking forward together, can we make better choices? (Extract from 'Memories and Seeds', written by Deborah Cobbett at the *Write About Time* workshop in Sheffield)

Introduction

Drawing on a unique data set, the previous chapters have discussed how residents of Jinja, Nanjing and Sheffield engage selectively with the 'global storm' and the 'intergenerational storm' of climate change. This chapter focuses on the INTERSECTION programme's complementary use of intergenerational community-based research and creative practice to respond to some of the themes and challenges raised. It includes three case studies: a *Write About Time* workshop led by Sheffield poet Helen Mort; participatory research to support environmental knowledge sharing in Jinja; and a *Sustainability Dancer* public artwork created by Sheffield sculptor Anthony Bennett. Through these case studies we explore the potential for creative practice to cultivate longitudinal thinking, provoke intergenerational dialogue and inspire sustainable action. To conclude, reflecting on findings from across the programme, we discuss what our research contributes to understandings of environmental justice.

Intergenerational community-based research and creative practice

The rationale for incorporating community-based, participatory and creative methodologies alongside conventional social research within the INTERSECTION programme was to explore the potential for coproduction, collaboration and arts practice to foster intergenerational dialogue. We were conscious of the ways in which, on the one hand, climate change is portrayed as a betrayal of younger generations by their elders and, on the other, younger generations are often blamed for unsustainable consumption practices. From the programme's inception through to implementation and analysis, we wanted to explore the potential of a mixed-methods approach for bringing people of different generations together in non-confrontational settings to generate new forms of environmental knowledge and action. This included an overlapping contingent of urban residents who participated in both interviews and intergenerational theatre workshops in each city, whose experiences are explored more extensively in the documentary film *Osbomb, love and*

Supershop: Performing sustainable worlds (2017). It also included the approaches that we discuss as case studies in this chapter. The value of this mixed-methods approach was that, at times, it enabled us to move beyond the production of research data toward a processual form of creative intergenerational practice that promoted improved intergenerational connections and facilitated meaningful action.

Our approach was informed by the ways in which social scientists have utilized forms of intergenerational practice to enrich their ethnographic understanding of particular contexts (Nordström, 2016), to creatively disseminate their research findings (Richardson, 2015), and to foster behaviour change (Kaplan and Haider, 2015). Hawkins (2015: 247) observes a recent creative 're-turn' that has seen social geographers increasingly working with experimental and 'art-full' methodologies that uphold 'geography's interdisciplinary relationship with arts and humanities', and notes 'the potential of creative methods for both researching and living differently'. In two of the case studies that we present here, creative and coproduced writing – including group and individual poems and short stories, some performed – demonstrates how research can retain the narrative authority of participants by considering 'whose voices get heard and which stories get told' (Mattingley, 2001: 44). The act of 'storying' experiences can foreground indigenous knowledge, foster intergenerational knowledge exchange, and form a bridge between the past and the present (Phillips and Bunda, 2018). Such acts, Haraway (1988; 1991) argues, enable people to create new meanings and the conditions under which alternative futures may be imagined, though it is also important to reflect critically on whose stories are told and heard. At its best, storying can help to address the need she identifies for: 'an earth wide network of connections, including the ability partially to translate knowledges among very different – power-differentiated – communities' (Haraway, 1988: 580). Our third case study is a public artwork that was produced in response to themes raised by our research participants, rather than coproduced with them. Here, visual art offers a means of expressing 'messy, unfinished and contingent' knowledge and bringing academic research to a wider public (Hawkins, 2015: 248).

Case study 1: Write About Time

A key challenge identified in Chapter Four's discussion of the moral geography of climate change was the relatively short-term 'generational timescape' within which most urban residents imagined responsibility to future generations, typically focused on their own lifetimes, their children and grandchildren. However, in Sheffield, our interest was kindled by a small number of residents who, when prompted to bring objects to signify 'something sustainable' or 'something you wish to save for future generations' to their interview, shared poems that for them expressed these ideas. These included Brian Patten's *So Many Different Lengths of Time*, W.B. Yeat's *Aedh Wishes for the Cloths of Heaven*, Mary Oliver's *Wild Geese* and Philip Larkin's *Going, Going*. These Sheffielders' choice of poems, while not explicitly focused on climate change, conveyed in various ways something of the 'deep time' that stretches beyond human lifespans (Girvan, 2014). This led to a collaboration with Sheffield poet Helen Mort, to explore whether creative writing might be of use in imagining more expansive timescapes.

Helen led a *Write About Time* workshop at Moor Theatre Delicatessen in Sheffield in July 2016, bringing around 30 people of different generations together to explore this theme through individual, paired and group writing prompts over the course of a day. Some participants had already taken part in interviews for the INTERSECTION programme, others would go on to do so, and some attended the workshop only. Some had a particular interest in creative writing, some the wider themes of the INTERSECTION programme, and others came out of curiosity. One of the participants, Sheffield blogger Ros Arksey, reflected afterwards that 'We were all here to write about time, to consider how it feels, how we would describe it and how others talk about time.'[1] Helen had suggested three poems for people to discuss in cross-generational groupings as a warm up activity: Michael Donaghy's *Upon a Claude Glass*, Kay Ryan's *The Edges of Time* and Percy Bysshe Shelley's *Ozymandias*. Together, these poems offered different perspectives on the fluidity of time, legacy, and looking forward and back. Ros noted that 'there was no right or wrong' in people's interpretations of these texts, and that the cross-generational groupings were interesting because 'some views were maybe influenced by where you were in your own life. For me this added richness to the discussion, as there were things that I didn't initially see or would not have thought about.'

Each person had been asked to bring something to the workshop that they would wish to include in a time capsule for future generations. Their first prompt was to work in cross-generational pairs to discuss their time capsule objects, and to each write one line to describe them. They were given the example of Craig Raine's poem *A Martian Sends a Postcard Home*, which renders the familiar strange in lines like 'Mist is when the sky is tired of flight/ and rests its soft machine on the ground'. This prompted them to think about how they might describe their objects to future generations who might not share the same frame of reference. The final exercise of the morning was to put these lines together in a *Sheffield Time Capsule* poem. Time capsule contributions ranged from practical objects ('a ladle serving food, a portal through which love is served') and advice on saving ('a procession of squirrels carrying sixpences and half crowns'[2]) to music ('it dances between major and minor') and

---

1 See Ros Arksey's blog, On time travelling: A reflection on the INTERSECTION Write About Time workshop', www.sheffield.ac.uk/intersection/news/writeabouttimeblog-1.595675.

2 This line references the design of a National Savings Stamp that the writer remembered from her childhood.

local produce ('a bottle of Henderson's relish'[3]). They included cherished places in the nearby countryside ('a deep pool of shimmering water, cold blue waves, lush green hills') and the post-industrial cityscape ('derelict walls, withered by rain and whispering ghosts'), as well as personal items such as a child's christening bracelet ('It is no bigger than the circle between my thumb and middle finger/ It fills with sky, with absence').

After lunch together, people were prompted to write to someone in the past or the future, with some imagining conversations with ex-boyfriends or lost relatives, others with distant strangers. There was also time for individual reflection and free writing responding to the themes of the workshop. The final section was an open mic, which commenced with a reading of the *Sheffield Time Capsule* poem followed by any other pieces written that day that people felt comfortable sharing. Afterwards, participants were given a deadline a month from the workshop to submit writing, if they wished, for publication on the INTERSECTION website. Their responses included reflections on the past ('a coal fire waiting for a match/ A hallway lined with faces pale as shells', Jim Caruth, 'What's Past'), generational experiences and legacies ('born in the NHS to orange juice and cod liver oil/All that Spirit of '45 stuff – the Trinity of Health, Housing and Education', Deborah Cobbett, 'Memories and Seeds'), the fluidity of time ('put your finger on it/ and hold it for a minute./ Here in a moment/ and then it retreats', Ros Arksey, 'Time'), and how a person's perspective on time can change ('I wonder, though, when I will/ begin to prefer/ yesterday to tomorrow', Urussa Malik, 'Presently').

These eloquent reflections on time illustrate how storying approaches can help to construct alternative timescapes, ones in which the past, present and future coexist in embodied and

[3] Henderson's Relish is a Sheffield-produced condiment similar to worcestershire sauce. The product's slogan is 'Made in Sheffield for over 100 years', so here the writer is making a joke about the product's longevity.

emplaced narratives (Phillips and Bunda, 2018). They suggest exciting possibilities for using storying to cultivate longitudinal thinking and broach questions about intergenerational obligations that proved difficult to imagine in a traditional interview setting. Girvan (2014: 363) suggests that 'storied spaces precede our entrance to the narrative and will succeed our exit from them', and that it is essential for climate change research to develop 'a deepening sense of storied time'. While this could only be explored to a limited extent in a one-day workshop, our next case study from Jinja highlights the benefits and impact of a more sustained coproduction process.

Case study 2: 'We Are the Foundations'

In Chapter Five's discussion of sustainable consumption, we highlighted Jinja residents' interconnected concerns about deforestation, waste management and the rising cost of charcoal. This led to the formation by some of our research participants of an intergenerational Green Briquette Cooperative to promote sustainable biomass consumption. This was Phase 2 of a creative intergenerational community-based participatory research process that began with a group of 11 volunteers drawn from We Are Walukuba, an intergenerational arts-for-development community organization founded through engagement with the INTERSECTION programme. The aim was to support intergenerational dialogue and knowledge sharing, bringing younger and older adults together on reciprocal terms to identify and address shared environmental concerns. Responding to themes emerging from our research, this process was envisaged as a way to bridge the older generation's experience of more subsistence-based ways of life and the younger generation's experience of urban livelihood insecurity and environmental degradation.

Phase 1 began with a series of cross-generational dialogues between the 11 participants (aged 19 to 56), about the ways in which the environment had transformed over generations. On the basis of these discussions, we worked with participants to co-design a short interview questionnaire. This focused on: what each generation had learned about protecting the environment; who they had learned this from; what role they believed their generation should play; and what skills they could share towards conserving the environment. Participants received basic training in qualitative research skills in workshops that employed discussion, roleplay and practice interviews to develop a shared understanding of the research process and guidelines for conducting ethical research. Each participant was then tasked with interviewing at least one person that they did not know from a different generation, and reporting their findings back to the group. Over the course of a week they conducted 27 interviews in total: 17 younger to older generation interviews and 10 older to younger generation interviews. These interviews were conducted in local languages, with responses recorded in notebooks and subsequently transcribed collectively into English.

Previous research has highlighted the potential for intergenerational environmental projects to raise awareness, increase participation, provide a focal point for strengthening intergenerational relationships and build community capacity (Kaplan and Liu, 2004; Wexler, 2011). Steinig and Butts (2009) suggest that, in such projects, older people play a particular role as 'mentors and guides'. This was reflected in the perceptions of the older people interviewed as part of this process, who saw themselves as 'the foundations' for environmental knowledge and action:

> 'The two generations must understand that their [grandparents] also lived and passed on, so they are living on their foundations and they also must implement on it for the future generations to find it.' (53-year-old man interviewed by 25-year-old man, Jinja)

Younger adults who had interviewed older people said they were surprised by how much they could learn from them, and discussed how the process had challenged negative stereotypes of older people's knowledge as "outdated" and "primitive":

> 'It was interesting to hear those stories of our parents. I couldn't believe what the old did in their time to protect the environment, the knowledge they used. I didn't have time to meet the older people, but now I have spent good time, I liked it. We talk, really freely. It was very cool, I liked it!' (Tony, male, late teens, Jinja)

While this demonstrates the potential for such projects to build relationships and solidarity across generations (Newman and Hattan-Yeo, 2008), there is a risk of inadvertently reinforcing characterizations of older people as repositories of knowledge and youth as receptacles (Tempest, 2003). Attempts to foreground equality and reciprocity between generations were also challenged by social norms associated with age and status. In practice, while the younger interviewers reported a positive experience of engaging with older generations, older interviewers – especially older women – struggled to find young people willing to talk to them, and reported being dismissed or laughed at in the attempt. Thus, the older generation occupied a paradoxical position as both an authority and a marginalized group within the power relations of the research process.

The group took their research back to We Are Walukuba to explore how they might use it to inform a creative performance to engage others in discussions about intergenerational cooperation and environmental sustainability. In making decisions about how to share what they had learned, and in reflecting on the negative experiences of older interviewers, they decided it was important to highlight the voices of the older generation in order to share their environmental knowledge. The local researcher and participants collaborated on writing a verbatim poem, 'We Are the Foundations', which was composed of direct quotations from interviews with older people in Jinja. The poem's themes include practical advice ('Wells made in swamps for water conservation/they would help during drought and combat famine'), indigenous beliefs ('The water bodies represented gods/ People were afraid to tamper with lakes, rivers and swamps/ Trees were protected/ in the belief they were inhabited by ancestors/ and if tampered with, demons would run/ and the place would soon become deserted'), and suggestions of how older and younger generations should work together to address shared environmental concerns ('We need to fight corruption/ by sensitising the government/ to prevent corrupt leaders/ from selling the wetlands to foreign investors'). It was performed by Samson (one of the younger participants) at a knowledge exchange workshop to an audience of representatives from Jinja Municipal Council, NGOs, local government and religious leaders, while an intergenerational group of six men and women embodied the words in a series of image theatre scenes. Afterwards, workshop participants were invited to reflect on the poem's themes and how they might inform 'strategies for creating a sustainable society for all ages'.

In subsequent discussions, the group and stakeholders identified deforestation and charcoal burning as key cross-generational environmental concerns. Participants continued to work together on creative waste action interventions and the formation of the Green Briquette Cooperative. A mutual respect developed between members of older and younger generations who began to see each other's potential as 'communities of knowledge' (Kuyken, 2012). Samson, in his mid-20s, felt that the group worked well together because "we do not stay with our skills, we are sharing and we teach them out and they teach other people". This case study and the ongoing work of the wider We Are Walukuba organization from which it is drawn illustrates how intergenerational community-based research and creative practice can, with sustained engagement, move beyond the production of research data to promote intergenerational cooperation and facilitate meaningful environmental action.

Case study 3: Sustainability Dancer

Our final case study is of a different kind of creative collaboration, one that sought to reflect on the cross-national and cross-generational themes raised over the course of the INTERSECTION programme. We commissioned Sheffield sculptor Anthony Bennett to create a public artwork to capture some of the key findings of our research, in order to stimulate reflection on how we might create a society for all ages and to remind us of our obligations to future generations. Anthony drew inspiration for this artwork from multiple conversations with the local researchers, as well as the INTERSECTION documentary film *Osbomb, love and Supershop: Performing sustainable worlds*, which showcases the use of intergenerational community theatre in each city. Anthony reflected afterwards:

> 'In Jinja, Nanjing and Sheffield INTERSECTION worked with different age groups, getting them to create theatrical tableaus addressing their concerns about sustainability and responsibility. The sculpture is my reaction to their concerns.'

The resulting artwork, a multimedia sculpture entitled *Sustainability Dancer*, was unveiled to Sheffield research participants, stakeholders and the general public at an event to mark the end of the project in March 2017. It is now on permanent display in the University of Sheffield's Faculty of Social Science, intended as a lasting legacy to provoke debate and reflection on global issues of sustainability, responsibility and justice.

In common with much of Anthony's work, the *Sustainability Dancer* is colourful, visually arresting and, on first viewing, somewhat of a puzzle. Anthony explained that he likes working with multiple visual details and themes, inviting people to question what it is they are seeing and why: "Each of the elements of this sculpture are like doorways… hopefully people will be intrigued to open one." The *Sustainability Dancer* is a ballerina balanced on-point, precariously, on a felled tree. Her gilded outstretched arms echo the universal figure of justice, traditionally depicted with a sword and scales, but here she carries a chainsaw and overflowing shopping bags. The chainsaw and various other elements of the sculpture, including her three-tiered tutu, appear to be made of cake. According to Anthony, the cake theme:

> 'came from thinking about intergenerational actions; times when generations come together – marriages, birthdays and christenings – when the focal point is always a cake, a celebration of their past, present and a hopeful future. But this cake is also a manifestation, and forewarning, of where we are in our world: a finely balanced global get together where some people get a small piece of the cake, some get a large piece.'

She wears a mask in reference in Nanjing residents' concerns about air pollution, a crown of upturned ice cream cones made from rolled-up currency, evoking waste, consumer excess and melting polar ice caps, and a seedling grows from the back of her head. This unwieldy figure balances on red shoes, in a nod to the folktale (and subsequent film and ballet) about the consequences of temptation.

Each tier of the tutu represents a different city, with particular details that come to the fore on repeat viewing. There are spoiled crops and parched earth in reference to peri-urban agriculture and drought in the Jinja layer at the base of the cake; fast food, electronics, consumer waste and overflowing jam in the Sheffield layer at the top of the cake; and in the Nanjing layer allusions to 'bling' culture, luxury goods and the Chinese parasol tree or *wútóng* (梧桐). At the base of the tree stump on which the *Sustainability Dancer* is balanced, accusatory fingers bloom from flowers or go around in circles. Also at the base is a quote from Yeats, one suggested by a Sheffield resident as encapsulating her idea of sustainability: 'Tread softly because you tread on my dreams'. A tie in the top layer represents the ubiquitous 'man in the suit', the powerful figures in industry and government who influence each city's development trajectory. Many of these visual motifs – the tie, the fast food, the seedling, the parasol trees, balloons in overflowing shopping bags, paper cut-outs – are drawn from theatrical scenes developed by city residents and appear in *Performing Sustainable Worlds.*

The *Sustainability Dancer* has much in common with 'a recent spate of artistic work focusing on (over)consumption using the lens of disposal and discard', in performing 'the old but powerful and necessary trick of taking something unthought and unseen and rendering it visible in new ways' (Crang, 2012: 763). It challenges those citizens represented in the top tier of the cake to think about the waste society creates and its global repercussions. It also confronts them with its messiness and multiplicity of meanings, intentionally overwhelming in its visual rendering of complex social research data from which there is no one unifying or universal story. This experimentation with visual art alongside storying approaches and intergenerational community theatre illustrates how creative methods 'are often situated as part of a tool-box of investigative techniques used by geographers and artists alike' (Hawkins, 2015: 263). Meanwhile, writers, sculptors, community theatre practitioners and research participants were able to use social inquiry to enrich their creative processes and speak to the key themes of the research in their own ways.

Concluding remarks: facing the future

The INTERSECTION programme has documented lived experiences of climate change, changing consumption practices and perspectives on intergenerational justice through the lens of the everyday lives of urban residents in three cities. It is, as far as we are aware, unique in its empirical focus on these issues, spanning cities in Europe, Africa and Asia. Much has been written about climate change and intergenerational justice from legal, philosophical and policy perspectives (for example, Barry, 1997; Page, 2006; McKinnon, 2011; Lawrence, 2015; Skillington, 2018), seeking to define what justice looks like and how it can be achieved. Such approaches have served as a locus for shaping political demands and encouraging policymakers to think of the long-term future. Movement between theory and practice is evident, for example, in the appointment of the Ombudsman for Future Generations in Hungary, the Future Generations Commissioner in Wales, parliamentary commissions in Israel and Finland, and moves within several countries to inscribe considerations of intergenerational justice within their constitutional arrangements (Fülöp, 2016: 198). These developments are welcome but, like sustainability discourses, predominantly 'top-down' and 'futures-orientated' (Agyeman et al, 2002). Middlemiss (2014: 942) identifies 'a need for sustainable development theory and policy to foster a nuanced understanding of the lived experiences of its subjects', suggesting that such approaches can benefit from complementary analyses of the connection between people's everyday experiences and the politics of climate change.

Our research highlights how, across Jinja, Nanjing and Sheffield, urban residents engage selectively with the 'global storm' and the 'intergenerational storm' of climate change. It highlights several important findings about the 'human sense of climate' (Hulme, 2017) that complicate moral readings of climate change as an international and intergenerational injustice.

Chapter Three on local narratives of climate change illustrated the social constructed-ness of climate change risk perception and the importance of local context (Bickerstaff and Walker, 2001; Hulme, 2017). Residents across Jinja, Nanjing and Sheffield were more or less anxious about climate change not only as a consequence of different levels of regional exposure, but also as a result of socioeconomic vulnerability to climate shocks, and the perceived physical deterioration or improvement of their immediate environment as a consequence of urban infrastructural change. Climate change, (de)industrialization and socioeconomic inequality are of course interrelated. However, the way they become conflated (or not) in the minds of urban residents looking for visible proof or disproof of climate change presents a real challenge to getting people living high consumption, low exposure lifestyles to recognize it as a serious threat. In particular, we found that residents in Sheffield were the least concerned about climate change, perceiving it as global, exotic, and remote in time and space.

Chapter Four on moral geographies of climate change explored how local perceptions of climate change intersect with considerations of environmental justice. We contrasted Jinja residents' narratives of self-blame for recent droughts, which linked (local) climate change with local causality, with Nanjing and Sheffield residents' focus on the global scale of climate change and 'meta-emitters' in government and industry. Each of these perspectives represents a partial diagnosis of the problem, and influences the extent to which people feel personally responsible for acting to address climate change. Significantly, we found that more people in Jinja felt that they have a major role to play in this regard than people living higher consumption lifestyles in Nanjing and Sheffield. This illustrates how it is possible to recognize the global injustice of climate change, and at the same time claim moral absolution as a minor player in a complex system, living with a kind of 'cognitive dissonance' by knowing and not acting (Klein, 2015: 3). Conversely, Jinja's residents' firm believe that their actions cause climate change irrespective of what foreign actors are doing highlights a double disparity in their exposure to environmental 'bads' and their access to information about global contributory factors.

Chapters Four and Five turned their attention to the intergenerational injustice of climate change, and the limited traction this has in everyday understandings of moral responsibility for action. We have illustrated how people tend to think about the future in the short term, focusing on their own children and grandchildren or on more immediate concerns to secure their own livelihood. This discourages consideration of how climate change is affecting others, now and in the more distant future. Moreover, people provide for themselves and their families in ways that they know to be unsustainable, indicating a moral challenge in reconciling responsibilities at different scales (Massey, 2004; Persson and Savulescu, 2012). We have also considered how arguments for climate change as 'intergenerational theft' (Nuccitelli, 2016) are undermined by a prevailing belief that younger generations consume more resources and live less sustainably than their elders. This complicates generational blame for climate change, illustrating how it intersects with other aspects of social change and intergenerational perceptions of (un)sustainable consumption. Given these divergent, coexisting narratives of generational blame, we have suggested that greater emphasis should be placed on intergenerational spaces and opportunities for dialogue about sustainability concerns.

Where climate change is perceived as more of an abstract moral issue affecting and enacted by distant strangers, attempts to engage citizens on an 'every little helps' basis premised on individual behaviour change are likely to prove ineffective (Shove, 2010; Phoenix et al, 2017). Conversely, blaming climate change on humanity's short-sightedness and local environmentally destructive practices served as a powerful motivator to act within a community on the frontline of drought, urban poverty and food insecurity. The case study from Jinja presented in this chapter illustrates how such concerns served as a locus for intergenerational dialogue and community-led action around environmental conservation (Eguavoen, 2013; Rudiak-Gould, 2014). This experience from Jinja is at once rousing and challenging to those in the West who fail to hold themselves – and the industries and governments they rightly identify as responsible for unsustainable development – accountable for the damage their lifestyles are inflicting on the planet.

In combining social research with applied arts methodologies, the INTERSECTION programme has demonstrated the potential for creatively sustaining engagement with communities and stakeholders on issues of sustainable development. In Jinja, our research has developed innovative models of good practice for citizen participation in sustainable development through community-based participatory research, knowledge exchange and co-designed and evidence-led strategies for sustainability. Through engaging in creative forms of knowledge coproduction and peer research over the course of 18 months, we have observed how the We Are Walukuba collective – a group of approximately 60 city residents drawn from a variety of ages and diversity of ethnic, educational and religious backgrounds – has overcome social barriers in the pursuit of social and environmental justice. Adopting a broad, community-led understanding of sustainability has enabled them to individually and collectively identify – and then seek to challenge – local intersections of social and environmental injustice. Intergenerational work in this way has generated new frames of engagement that seek to enhance inclusivity and generate transformative action (see McQuaid and Plastow, 2017).

Data collected from interviews, dialogue groups and applied theatre processes were used to coproduce creative forms of knowledge exchange with Jinja's residents. Through participating in a series of creative events, community members have played an active role in engaging with wider publics and policymakers on issues of critical importance to them, including deforestation, gender-based violence and waste management. At the workshop mentioned earlier in this chapter, for example, roundtable discussions were facilitated between community members and politicians, Jinja Municipal Council staff, technical officers including town planners and environment officers, and cultural and religious leaders. The re-presentation of research findings through creative practice provided new mechanisms for making visible entrenched environmental and social issues. Policymakers and practitioners participating in these events gained better understanding of community priorities, perspectives and contexts for change, creating opportunities to better relate to and serve the communities in question.

While in no way wishing to detract from the substantial achievements of residents in Jinja and the We Are Walukuba collective they formed as a result of their engagement in this research, it is interesting to note that intergenerational community-based research in Nanjing and Sheffield did not result in or sustain anything like the same level of momentum. We could reflect on this in itself as an environmental justice issue, with the least polluting and least well-resourced group of residents seemingly the most willing to organize collectively to act. Middlemiss (2014) reflects on the extent to which the advent of consumer society creates individualized subjects and undermines civic participation, suggesting that poorer, less polluting people may be more predisposed to participate in sustainable development initiatives. It is also important to critically reflect on the politics of encouraging citizens to take personal and communal responsibility for environmental 'goods' and 'bads' through community-based participatory research. In particular, the extent to which this may inadvertently reinforce neoliberal approaches to the management of nature (Castree, 2008), and/or empower communities to secure justice and recognition by improving access to environmental information, inclusion in policy and decision-making processes, and access to legal processes for protecting environmental rights (Walker, 2011; Agyeman, 2013).

Altogether, these accounts provide evidence for the significance of local narratives and moralities of climate change in mobilizing publics to act to secure environmental justice. It is not simply a case of providing the 'right' information from a geoscientific perspective, but also understanding lived experiences of climate vulnerability, environmental degradation and regeneration, which, we have argued, have particular characteristics in urban (post)industrial contexts. The lens of 'political responsibility' for climate change, which focuses on shared culpability, collective action and caring at a distance by virtue of our connectedness, is useful for engaging with people who view it as remote in space and time, or else feel overwhelmed by its global and intergenerational complexities.

References

Abbink, G.J. (2005) 'Being young in Africa: The politics of despair and renewal', in G. Abbink and I. van Kessel (eds) *Vanguard or vandals: Youth, politics and conflict in Africa*, Leiden: Brill, pp 1–36.

Agyeman, J. (2000) *Environmental justice: From the margins to the mainstream?,* London: Town and Country Planning Association.

Agyeman, J. (2013) *Introducing just sustainabilities: Policy, planning and practice*, London: Zed Books.

Agyeman, J., Bullard, R.D. and Evans, B. (2002) 'Exploring the nexus: Bringing together sustainability, environmental justice and equity', *Space and Polity* 6(1): 77–90.

Akwango, D.A., Obaa, B.B., Turyahabwe, N., Baguma, Y. and Egeru, A. (2016) 'Agro-pastoral choice of coping strategies and response to drought in the semi-arid areas of Uganda', *African Journal of Rural Development*, 1(3): 281–91.

Appadurai, A. (2004) 'The capacity to aspire: Culture and terms of recognition', in V. Rao and M. Walton (eds.) *Culture and public action*, Washington, DC: The World Bank and Stanford University Press, pp 59–84.

Ariztia, T., Kleine, D., Bartholo, R., Brightwell, G., Agloni, N. and Afonso, R. (2016) 'Beyond the "deficit discourse": Mapping ethical consumption discourses in Chile and Brazil', *Environment and Planning A*, 48(5): 891–909.

Attas, D. (2009) 'A transgenerational difference principle', in A. Gosseries and L.H. Meyer (eds) *Intergenerational justice*, Oxford: Oxford University Press, pp 189–218.

Baatz, C. (2013) 'Responsibility for the past? Some thoughts on compensating those vulnerable to climate change in developing countries', *Ethics, Policy & Environment*, 16(1): 94–110.

Bailey, I. and Wilson, G. (2009) 'Theorising transitional pathways in response to climate change: Technocentrism, ecocentrism, and the carbon economy', *Environment and Planning A*, 41(10): 2324–41.

Banks, N. (2015) 'Understanding youth: Towards a psychology of youth poverty and development in sub-Saharan African cities', Brooks World Poverty Institute Working Paper Series No. 216.

Barnes, J. and Dove, M.R. (2015) 'Introduction', in J. Barnes and M.R. Dove (eds) *Climate cultures: Anthropological perspectives on climate change*, New Haven and London: Yale University Press, pp 1–24.

Barnett, C., Cloke, P., Clarke, N. and Malpass, A. (2005) 'Consuming ethics: Articulating the subjects and spaces of ethical consumption', *Antipode*, 37(1): 23–45.

Barnett, C., Cloke, P., Clarke, N. and Malpass, A. (2010) *Globalizing responsibility: The political rationalities of ethical consumption*, Chichester: Wiley-Blackwell.

Barr, S., Shaw, G. and Coles, T. (2011) 'Sustainable lifestyles: Sites, practices, and policy', *Environment and Planning A*, 43(12): 3011–29.

Barry, B. (1997) 'Sustainability and intergenerational justice', *Theoria*, 89: 43–64.

Barry, J. (2003) 'Ecological modernisation', in E. Page and J. Proops (eds) *Environmental thought*, Cheltenham: Edward Elgar, pp 191–213.

Bashaasha, B., Thomas, T.S., Waithaka, M. and Kyotalimye, M. (2013) 'Uganda', in M. Waithaka, G.C. Nelson, T.S. Thomas, and M. Kyotalimye (eds) *East African agriculture and climate change: A comprehensive analysis*, Washington, DC: International Food Policy Research Institute, pp 347–87.

BBC News (2019) 'Climate strike: schoolchildren protest over climate change', *BBC News online*, 15 February 2019, www.bbc.co.uk/news/uk-47250424

BEIS (Department for Business, Energy and Industrial Strategy) (2017) *Final UK greenhouse gas emissions national statistics: 1990–2017*, https://www.gov.uk/government/statistics/final-uk-greenhouse-gas-emissions-national-statistics-1990-2017

Berners-Lee, M. and Clark, D. (2013) *The burning question: We can't burn half the world's oil, coal and gas, so how do we quit?* London: Profile Books.

Best, K. (2014) 'Diaosi: China's "loser" phenomenon', *On Politics*, 7(1): 20–31.

Bialasiewicz, L. and Minca, C. (2005) 'Old Europe, new Europe: For a geopolitics of translation', *Area*, 37(4): 365–72.

Bickerstaff, K. and Walker, G. (2001) 'Public understandings of air pollution: The "localisation' of environmental risk"', *Global Environmental Change*, 11(2): 133–45.

Birdwhistell, J.D. (2007) *Mencius and masculinities: Dynamics of power, morality, and maternal thinking*, New York: State University of New York Press.

Boersch-Supan, J. (2012) 'The generational contract in flux: Intergenerational tensions in post-conflict Sierra Leone', *The Journal of Modern African Studies*, 50(1): 25–51.

Borchgrevink, A. (2003) 'Silencing language: Of anthropologists and interpreters', *Ethnography*, 4(1): 95–121.

Brace, C. and Geoghegan, H. (2010) 'Human geographies of climate change: Landscape, temporality and lay knowledges', *Progress in Human Geography*, 35(3): 284–302.

Bullard, R. (1990) *Dumping in Dixie: Race, class and environmental quality*, Boulder, CO: Westview Press.

Bulmer, M. (1998) 'The problem of exporting social survey research', *American Behavioral Scientist*, 42(2): 153–67.

Burgess, T. and Burton, A. (2010) 'Introduction', in A. Burton and H. Charton-Bigot (eds) *Generations past: Youth in East African history*, Athens, OH: Ohio University Press, pp 1–24.

Burke, E. and Mitchell, L.G. (2009 [1790]) *Reflections on the revolution in France* (Oxford World's Classics edition), Oxford: Oxford University Press.

Cameron, L., Erkal, N., Gangadharan, L. and Meng, X. (2013) 'Little Emperors: Behavioural impacts of China's one-child policy', *Science*, 339(6122): 953–7.

Caney, S. (2005) 'Cosmopolitan justice, responsibility and global climate change', *Leiden Journal of International Law*, 18(4): 747–75.

Capstick, S., Whitmarsh, L., Poortinga, W., Pidgeon, N. and Upham, P. (2015) 'International trends in public perceptions of climate change over the past quarter century', *WIREs Climate Change*, 6(1): 35–61.

Carr, J., Gotlieb, M.R., Lee, N. and Shah, D.V. (2012) 'Examining overconsumption, competitive consumption, and conscious consumption from 1994 to 2004: Disentangling cohort and period effects', *The Annals of the American Academy of Political and Social Science*, 644(1): 220–33.

Castree, N. (2008) 'Neoliberalising nature: The logics of deregulation and reregulation', *Environment and Planning A*, 40(1): 131–52.

Castree, N. (2017) 'Global change research and the "people's disciplines": Toward a new dispensation', *South Atlantic Quarterly*, 116(1): 55–67.

Chan, R.Y.K. and Lau, L.B.Y. (2000) 'Antecedents of green purchases: A survey in China', *Journal of Consumer Marketing*, 17(4): 338–57.

Christiansen, C. Utas, M. and Vigh, H. (2006) 'Youth(e)scapes', in C. Christiansen, M. Utas and H.E. Vigh (eds) *Navigating youth generating adulthood: Social becoming in an African context*, Uppsala: Nordiska Afrikainstitutet, pp 9–28.

Christophers, B. (2017) 'Intergenerational inequality? Labour, capital and housing through the ages', *Antipode*, 50(1): 101–21.

Collard, D. (2000) 'Generational transfers and the intergenerational bargain', *Journal of International Development*, 12(4): 453–62.

Committee on Climate Change (2016) *UK climate change risk assessment 2017: Evidence report*, London: Committee on Climate Change.

Cooper, E. and Pratten, D. (2015) 'Ethnographies of uncertainty in Africa: An introduction', in E. Cooper and D. Pratten (eds) *Ethnographies of uncertainty in Africa*, Basingstoke: Palgrave Macmillan, pp 1–16.

Crang, M. (2012) 'Negative images of consumption: Cast offs and casts of self and society', *Environment and Planning A: Economy and Space*, 44(4): 763–7.

Cuomo, C.J. (2011) 'Climate change, vulnerability and responsibility', *Hypatia*, 26(4): 690–714.

Dalton, P.S., Ghosal, S. and Mani, A. (2016) 'Poverty and aspirations failure', *The Economic Journal*, 126(590): 165–88.

Darier, E. and Schule, R. (1999) 'Think globally, act locally? Climate change and public participation in Manchester and Frankfurt', *Local Environment*, 4(3): 317–29.

DEFRA (Department for Environment, Food and Rural Affairs) (2002) *Survey of public attitudes to quality of life and to the environment: 2001*, London: DEFRA.

DEFRA (2007) *Survey of public attitudes and behaviors towards the environment: 2007*, London: DEFRA.

DEFRA (2015) *Statistics on waste management by local authorities in England in 2014-2015*, www.gov.uk/government/uploads/system/uploads/attachment_data/file/481771/Stats_Notice_Nov_2015.pdf

Den Elzen, M. and Schaeffer, M. (2002) 'Responsibility for past and future global warming: Uncertainties in attributing anthropogenic climate change', *Climatic Change*, 54(1-2): 29–73.

Dermody, J., Hanmer-Lloyd, S., Koenig-Lewis, N. and Zhao, A.L. (2015) 'Advancing sustainable consumption in the UK and China: The mediating effect of pro-environmental self-identity', *Journal of Marketing Management*, 31(13–14): 1472–502.

Desbiens, C. and Ruddick, S. (2006) 'Guest editorial: Speaking of geography: Language, power, and the spaces of Anglo-Saxon "hegemony"', *Environment and Planning D: Society and Space*, 24(1): 1–8.

Dewilde, C. (2003) 'A life-course perspective on social exclusion and poverty', *British Journal of Sociology*, 54(1): 109–28.

Dickinson, J.L. (2009) 'The people paradox: Self-esteem striving, immortality ideologies, and human response to climate change', *Ecology and Society*, 14, unpaginated.

Dieter, G. and Bergmann, S. (2011) *Religion in environmental and climate change: Suffering, values, lifestyles*, Oxford: Bloomsbury.

Diprose, K., Valentine, G., Vanderbeck, R.M., Liu, C. and McQuaid, K. (2019a) 'Building common cause towards sustainable consumption: A cross-generational perspective', *Environment and Planning E: Nature and Space*, 2(2) 203–228.

Diprose, K., Liu, C., Valentine, G., Vanderbeck, R.M. and McQuaid, K. (2019b) 'Caring for the future: Climate change and intergenerational responsibility in China and the UK', *Geoforum*, advanced online publication, https://doi.org/10.1016/j.geoforum.2019.05.019

Dodman, D. and Satterthwaite, D. (2009) 'Institutional capacity, climate change adaptation and the urban poor', *IDS Bulletin*, 39(4): 67–74.

Duflo, E. (2013) 'Hope, aspirations and the design of the fight against poverty', Arrow Lectures, Stanford University, https://ethicsinsociety.stanford.edu/events/esther-duflo-hope-aspirations-and-design-fight-against-poverty

Dolan, P. (2002) 'The sustainability of "sustainable consumption"', *Global Policy and Environment*, 22(2): 170–81.

Durham, D. (2000) 'Youth and the social imagination in Africa: Introduction to parts 1 and 2', *Africa Quarterly*, 73(3): 113–20.

Dwyer, J.F., Schroeder, H.W. and Gobster, P.H. (1991) 'The significance of urban trees and forests: Towards a deeper understanding of values', *Journal of Arboriculture*, 17(10): 276–84.

Edelman, L. (2004) *No future: Queer theory and the death drive*, Durham, NC: Duke University Press.

Eguavoen, I. (2013) 'Climate change and trajectories of blame in Northern Ghana', *Anthropological Notebooks*, 19(1): 5–24.

Evans, D. (2011a) 'Beyond the throwaway society: Ordinary domestic practice and a sociological approach to household food waste', *Sociology*, 46(1): 41–56.

Evans, D. (2011b) 'Thrifty, green or frugal: Reflections on sustainable consumption in a changing economic climate', *Geoforum*, 42(5): 550–7.

Evans, D., Welch, D. and Swaffield, J. (2017) 'Constructing and mobilizing 'the consumer': Responsibility, consumption and the politics of sustainability', *Environment and Planning A*, 49(6): 1396–412.

Evans, R. (2011) '"We are managing our own lives...": Life transitions and care in sibling-headed households affected by AIDS in Tanzania and Uganda', *Area*, 43(4): 384–96.

Evans, R. (2012) 'Sibling caringscapes: Time-space practices of caring within youth-headed households in Tanzania and Uganda', *Geoforum*, 43(4): 824–35.

Ewing, J.A. (2017) 'Hollow ecology: Ecological modernization theory and the death of nature', *Journal of World Systems Research*, 23(1): 126–55.

Ferguson, H. (2012) *Briquette businesses in Uganda: The potential for briquette enterprises to address the sustainability of the Ugandan biomass fuel market*, London: GIVAP International.

Fincher, R., Barnett, J., Graham, S. and Hurlimann, A. (2014) 'Time stories: Making sense of futures in anticipation of sea level rise', *Geoforum*, 56: 201–10.

Forkert, K. (2014) 'The new moralism: Austerity, silencing and debt morality', *Soundings*, 56: 41–53.

Frederiksen, B.F. and Munive, J. (2010) 'Young men and women in Africa: Conflicts, enterprise and aspiration', *Young*, 18(3): 249–58.

Fu, Y. (2005) 'Intergenerational justice and harmony from a traditional perspective', *Journal of Hunan University of Science and Engineering*, 26(7): 50–2. [符艳红.2005.从中国传统文化视角论代际公平与和谐.湖南科技学院学报, 26 (7), 50–2.]

Fu, Y. (2007) 'The humanistic approach of 'Intergenerational justice and harmony'', *Qiu Suo*, 1: 138–40. [符艳红 (2007) "代际公平与和谐"的人性基础初探. 求索, 1, 138–40.]

Fülöp, S. (2016) 'The institutional representation of future generations', in G. Bos and M. Düwel, (eds), *Human rights and sustainability: Moral responsibilities for the future*, New York: Routledge/Taylor & Francis Group, pp 195–211.

Füssel, H-M. (2010) 'How inequitable is the global distribution of responsibility, capability and global vulnerability to climate change: A comprehensive indicator based assessment', *Global Environmental Change*, 20(4): 597–611.

Gaard, G. (2015) 'Ecofeminism and climate change', *Women's Studies International Forum*, 49: 20–33.

Gardiner, S.M. (2001) 'The real tragedy of the commons', *Philosophy & Public Affairs*, 30(4): 387–416.

Gardiner, S.M. (2006) 'A perfect moral storm: Climate change, intergenerational ethics and the problem of moral corruption', *Environmental Values*, 15(3): 397–413.

Gatens, M. and Lloyd, G. (1999) *Collective imaginings: Spinoza, past and present*, London: Routledge.

Girvan, A. (2014) 'Cultivating longitudinal knowledge: Alternative stories for an alternative chronopolitics of climate change', in R. Boschman and M. Trono (eds) *Found in Alberta: Environmental themes in the Anthropocene*, Waterloo Ontario: Wilfrid Laurier University Press, pp 347–70.

Green Climate Fund (2018) 'About the fund', www.greenclimate.fund/who-we-are/about-the-fund

Gregson, N. and Crang, M. (2010) 'Materiality and waste: Inorganic vitality in a networked world', *Environment and Planning A*, 42(5): 1026–32.

Gregson, N., Metcalfe, A. and Crewe, L. (2007) 'Identity, mobility and the throwaway society', *Environment and Planning D*, 25(4): 682–700.

Goodman, D.S.G. (2014) 'Middle class China: Dreams and aspirations', *Journal of Chinese Political Science*, 19(1): 49–67.

Guyer, J. (2007) 'Prophecy and the near future: Thoughts on macroeconomic, evangelical, and punctuated time', *American Ethnologist*, 34(3): 409–21.

Hackmann, H., Moser, S.C and St. Clair, A.L. (2014) 'The social heart of global environmental change', *Nature Climate Change*, 4(7): 653–5.

Hamilton, C. (2010) *Requiem for a species: Why we resist the truth about climate change*, London: Earthscan.

Hamilton, C., Bonneuil, C. and Gemenne, F. (2015) 'Thinking the Anthropocene', in C. Hamilton, C. Bonneuil and F. Gemenne (eds) *The Anthropocene and the global environmental crisis: Rethinking modernity in a new epoch*, London: Routledge, pp. 1–15.

Hammond, P. (2017) *Autumn Budget 2017: Philip Hammond's speech*, 22 November 2017, www.gov.uk/government/speeches/autumn-budget-2017-philip-hammonds-speech

Hansen, J. and Sato, M. (2016) 'Regional climate change and national responsibilities', *Environmental Research Letters*, 11(3): 1–9.

Haraway, D.J. (1988) 'Situated knowledge: The science question in feminism and the privilege of partial perspective', *Feminist Studies*, 14(3): 575–99.

Haraway, D.J. (1991) *Simians, cyborgs, and women: The reinvention of nature*, Abingdon: Routledge.

Harley, T.A. (2003) 'Nice weather for the time of year: The British obsession with weather', in S. Strauss and B.S. Orlove (eds) *Weather, climate, culture,* London: Berg, pp 103–18.

Harzing, A.-W. (2006) 'Response styles in cross-national survey research: A 26-country study', *International Journal of Cross Cultural Management,* 6: 243–66.

Hawkins, H. (2015) 'Creative geographic methods: Knowing, representing, intervening. On composing page and place', *Cultural Geographies*, 22(2): 247–68.

Hayhurst, L. (2014) 'The 'Girl Effect' and martial arts: Social entrepreneurship and sport, gender and development in Uganda', *Gender, Place & Culture*, 21(3): 297–315.

He, J. and van de Vijver, F. (2012) 'Bias and equivalence in cross-cultural research', *Online Readings in Psychology and Culture*, 2(2).

Henley, R. (2010) 'Resilience enhancing psychosocial programmes for youth in different cultural contexts: Evaluation and research', *Progress in Development Studies*, 10(4): 295–307.

Hillier, A. (2011) 'Climate change and individual responsibility', *The Monist*, 94(3): 349–368.

Hobson, K. (2002) 'Competing discourses of sustainable consumption: Does the 'rationalisation of lifestyles' make sense?', *Environmental Politics*, 11(2): 95–120.

Holdaway, J. (2010) 'Environment and health in China: An introduction to an emerging research field', *Journal of Contemporary China*, 19(63): 1–22.

Honwana, A. (2012) *The time of youth: Work, social change, and politics in Africa*, Washington, DC: Kumarian Press.

House of Commons Environmental Audit Committee (2017) *Sustainable development goals in the UK*, London: House of Commons.

House of Commons Work and Pensions Committee (2016) *Intergenerational Fairness: Third report of session 2016–2017*, London: House of Commons.

Howker, E. and Malik, S. (2010) *Jilted generation: How Britain has bankrupted its youth*, London: Icon Books.

Hu, Y., Hasan, N., Wang, S., Zhou, Y., Yang, T. and Zhang, Y. (2017) 'Implicit and explicit attitudes of Chinese youth toward the second-generation rich', *Social Behavior and Personality: An International Journal*, 45(3): 427–40.

Hulme, M. (2011) 'Reducing the future to climate: A story of climate determinism and reductionism', *Osiris*, 26(1): 245–66.

Hulme, M. (2017) *Weathered: Cultures of climate*, London: Sage.

Hulme, M., Dessai, S., Lorenzoni, I. and Nelson, D.R. (2009) 'Unstable climates: Exploring the statistical and social constructions of 'normal science'', *Geoforum*, 40(2): 197–206.

Inglehart, R. (1971) 'The silent revolution in Europe: Intergenerational change in post-industrial societies', *American Political Science Review*, 65(4): 991–1017.

Inglehart, R.F. (2008) 'Changing values among Western publics from 1970 to 2006', *Western European Politics*, 31(1-2): 130–46.

International Energy Agency (2016) *Key world energy statistics 2016*, Paris: IEA.

IPCC (Intergovernmental Panel on Climate Change) (2014) *Climate change 2014: Synthesis report. Contribution of Working Groups I, II and III to the fifth assessment report of the Intergovernmental Panel on Climate Change*. R.K. Pachauri and L.A. Meyer (eds), Geneva, Switzerland: IPCC.

IPCC (2018) *Global warming of 1.5°C: An IPCC Special Report on the impacts of global warming of 1.5°C above pre-industrial levels and related global greenhouse gas emission pathways, in the context of strengthening the global response to the threat of climate change, sustainable development, and efforts to eradicate poverty*, Geneva, Switzerland: IPCC.

Jackson, P., Ward, N. and Russell, P. (2009) 'Moral economies of food and geographies of responsibility', *Transactions of the Institute of British Geographers*, 34(1): 12–24.

Jackson, R.B., Le Quéré, C., Andrew, R.M., Canadell, J.G., Korsbakken, J.I., Liu, Z., Peters, G.P. and Zheng, B. (2018) 'Global energy growth is outpacing decarbonization', *Environment Research Letters*, 13(12): 1–7.

Jackson, T. (2005) 'Live better by consuming less? Is there a "double dividend" in sustainable consumption?', *Journal of Industrial Ecology*, 9(1–2): 19–36.

Jackson, T. (2009) *Prosperity without growth: Economics for a finite planet*, London: Earthscan.

Jamieson, D. (2010) 'Climate change, responsibility and justice', *Science and Engineering Ethics*, 16(3): 431–45.

Jensen, T. (2013) 'Riots, restraint and the new cultural politics of wanting', *Sociological Research Online*, 18(4): 1–12.

Jing, J. (2000) *Feeding China's Little Emperors: Food, children, and social change*, Stanford: Stanford University Press.

Johnson, T.P. (1998) 'Approaches to equivalence in cross-cultural and cross-national surveys', *ZUMA Nachrichten Spezial: Cross-Cultural Survey Equivalence*, 3: 1–40.

Johnson, T.P. and van de Vijver, F.J.R. (2003) 'Social desirability in cross-cultural research', in J.A. Harkness, F.J.R. van de Vijver and P.P. Mohler (eds), *Cross-cultural survey methods*, New York: Wiley, pp 195–204.

Johnson-Hanks, J. (2005) 'When the future decides: Uncertainty and intentional action in contemporary Cameroon', *Current Anthropology*, 46(3): 363–85.

Jones, E.L. (1963) 'The courtesy bias in South-East Asian surveys', *International Social Science Journal*, 15: 70–6.

Kan, K. (2013) 'The new "lost generation": Inequality and discontent among Chinese youth', *China Perspectives*, 2013/2: 63–73.

Kaplan, M. and Haider, J. (2015) 'Creating intergenerational spaces that promote health and well-being', in R.M. Vanderbeck and N. Worth (eds) *Intergenerational space*, London: Routledge, pp 33–49.

Kaplan, M.S. and Liu, S. (2004) *Generations United for environmental awareness and action*, Washington, DC: Generations United.

Keillor, B., Owens, D. and Pettijohn, C. (2001) 'A cross-cultural/cross-national study of influencing factors and socially desirable response biases', *International Journal of Market Research*, 43: 63–84.

Keith, M., Lash, S., Arnoldi, J. and Rooker, T. (2014) *China constructing capitalism: Economic life and urban change*, London: Routledge.

Klein, E. (2014) 'Psychological agency: Evidence from the urban fringe of Bamako', *World Development*, 64: 642–53.

Klein, N. (2015) *This changes everything: Capitalism vs. the climate*, New York: Simon and Schuster.

Kong, L. (2010) 'China and geography in the 21st century: A cultural (geographical) revolution?', *Eurasian Geography and Economics*, 51(5): 600–18.

Kuyken, K. (2012) 'Knowledge communities: Towards a re-thinking of intergenerational knowledge transfer', *VINE: The Journal of Information and Knowledge Management Systems*, 42(3/4): 365–81.

Langevang, T. (2008) '"We are managing!" Uncertain paths to respectable adulthoods in Accra Ghana', *Geoforum*, 39(6): 2039–47.

Lawrence, P. (2015) *Justice for future generations: Climate change and international law*, Cheltenham: Edward Elgar Publishing.

Leiserowitz, A., Maibach, E., Roser-Renouf, C., Rosenthal, S. and Cutler, M. (2017) *Climate change in the American mind: May 2017*, New Haven, CT: Yale University and George Mason University.

Le Quéré, C. et al (2017) 'Global Carbon Budget 2017', *Earth System Science Data*, 10(1): 405–48.

Lewis, T.L. (2000) 'Media Representations of "Sustainable Development"', *Science Communication,* 21(3): 244–73.

Li, B. and Shin, H.B. (2013) 'Intergenerational housing support between retired old parents and their children in urban China', *Urban Studies*, 50(16): 3225–42.

Li, X. and Tilt, B. (2017) 'Perceptions of quality of life and pollution among China's urban middle class: The case of smog in Tangshan', *China Quarterly*, 234: 340–56.

Li, Y., Johnson, E.J. and Zaval, L. (2011) 'Local warming: Daily temperature change influences belief in global warming', *Psychological Science*, 22(4): 454–9.

Liao, X. and Cheng, H. (2004) 'On intergenerational justice', *Ethics Research*, 4.

Little, B. and Winch, A. (2017) 'Generation: The politics of patriarchy and social change', *Soundings*, 66: 129–44.

Liu, C., Chen. L., Vanderbeck, R.M., Valentine, G., Zhang, M., Diprose, K. and McQuaid, K. (2018) 'A Chinese route to sustainability: Postsocialist transitions and the construction of ecological civilisation', *Sustainable Development*, 26(6): 741–8, doi.org/10.1002/sd.1743.

Liu, C., Valentine, G., Vanderbeck, R.M., McQuaid, K. and Diprose, K. (2018) 'Placing " sustainability" context: Narratives of sustainable consumption in Nanjing, China', *Social and Cultural Geography*, doi.org/10.1080/14649365.2018.1454978.

Liu, J. (2014) 'Ageing, migration and familial support in rural China', *Geoforum*, 51: 305–12.

Liu, W. (2017) 'Intergenerational emotion and solidarity in transitional China: Comparisons of two kinds of "ken lao" families in Shanghai', *The Journal of Chinese Sociology*, 4(10): 1–22.

Lohr, V.I., Pearson-Mims, C.H., Tarnal, J. and Dillman, D.A. (2004) 'How urban residents rate and rank the benefits and problems associated with trees in cities', *Journal of Arboriculture*, 30(1): 28–35.

Lorenzoni, I., Nicholson-Cole, S. and Whitmarsh, L. (2007) 'Barriers perceived to engaging with climate change among the UK public and their policy implications', *Global Environmental Change*, 17(3–4): 445–59.

Ma, C.B. (2010) 'Who bears the environmental burden in China? An analysis of the distribution of industrial pollution sources', *Ecological Economics*, 69(9): 1869–76.

Magrath, J. (2008) *Turning up the heat: Climate change and poverty in Uganda*, Kampala/Oxford: Oxfam GB.

Mahony, M. and Hulme, M. (2016) 'Epistemic geographies of climate change: Science, space and politics', *Progress in Human Geography*, 42(3): 395–424.

Mains, D. (2012) 'Blackouts and progress: Privatization, infrastructure, and a developmentalist state in Jimma, Ethiopia', *Cultural Anthropology*, 27(1): 3–27.

Maniates, M.F. (2001) 'Individualization: Plant a tree, buy a bike, save the world?', *Global Environmental Politics*, 1(3): 31–52.

Mannheim, K. (1952 [1923]) 'The problem of generations', in K. Mannheim, *Essays on the sociology of knowledge*, London: RKP.

Marshall, G. (2014) *Don't even think about it: Why our brains are wired to ignore climate change*, London: Bloomsbury.

Massey, D. (2004) 'Geographies of responsibility', *Geographiska Annaler: Series B, Human Geography*, 86(1): 5–18.

Massey, D. (2007) *World city*, Oxford: Polity.

Mattingley, D. (2001) 'Place, teenagers and representations: Lessons from a community theatre project', *Social and Cultural Geography*, 2(4): 445–59.

Mbembe, A. (2002) 'African modes of self writing', *Public Culture*, 14(1): 239–73.

McDonald, C. (2015) 'How many Earths do we need?' *BBC News Magazine*, 16 June, www.bbc.co.uk/news/magazine-33133712

McDonald, R.I., Green, P., Balk, D., Fekete, B.M., Revenga, C., Todd, M. and Montgomery, M. (2011) 'Urban growth, climate change, and freshwater availability', *Proceedings of the National Academy of Sciences*, 108(15): 6312–17.

McEwan, C. and Goodman, M.K. (2010) 'Place geography and the ethics of care: Introductory remarks on the geographies of ethics, responsibility and care', *Ethics, Place and Environment*, 13(2): 103–12.

McKibben, B. (2003 [1989]) *The end of nature: Humanity, climate change and the natural world*, 2nd revised edition, London: Bloomsbury.

McKibben, B. (2012) 'Global warming's terrifying new math', *Rolling Stone*, 19 July, www.rollingstone.com/politics/politics-news/global-warmings-terrifying-new-math-188550/

McKinnon, C. (2011) *Climate change and future justice*, London: Routledge.

McLaren, D. (2003) 'Environmental space, equity and the ecological debt', in J. Agyeman, R.D. Bullard and B. Evans (eds) *Just sustainabilities: Development in an unequal world*, London: Earthscan, pp 19–37.

McQuaid, K. and Plastow, J. (2017) 'Ethnography, applied theatre and stiwanism: Creative methods in search of praxis amongst men and women in Jinja, Uganda', *Journal of International Development*, 29(7): 961–80.

McQuaid, K., Vanderbeck, R.M., Plastow, J., Valentine, G., Liu, C., Chen, L., Zhang, M. and Diprose, K. (2017) 'Intergenerational community-based research and creative practice: Promoting environmental sustainability in Jinja, Uganda', *Journal of Intergenerational Relationships*, 15(4): 389–410.

McQuaid, K., Vanderbeck, R.M., Valentine, G., Liu, C., Chen L., Zhang, M. and Diprose. K. (2018a) 'Urban climate change, livelihood vulnerability and narratives of generational responsibility in Jinja, Uganda', *Africa*, 88(1): 11–37.

McQuaid, K., Vanderbeck, R.M., Valentine, G., Liu, C. and Diprose, K. (2018b) 'An elephant cannot fail to carry its own ivory': Transgenerational ambivalence, infrastructure and sibling support practices in urban Uganda', *Emotion, Space and Society*, doi.org/10.1016/j.emospa.2018.07.009.

MEMD (Ministry of Energy and Mineral Development) (2013) *Strategic Investment Plan 2014/15 – 2018/19*. Kampala, Uganda: Government of Uganda.

Meze-Hausken, E. (2004) 'Contrasting climate variability and meteorological drought with perceived drought and climate change in northern Ethiopia', *Climate Research*, 27(1): 19–31.

Middlemiss, L. (2014) 'Individualised or participatory? Exploring late-modern identity and sustainable development', *Environmental Politics*, 23(6): 929–46.

Moser, S. and Kleinhückelkotten, S. (2018) 'Good intents, but low impacts: Diverging importance of motivational and socioeconomic determinants explaining pro-environmental behaviour, energy use, and carbon footprint', *Environment and Behaviour*, 50(6): 626–56.

Mukiibi, J.K. (2001) *Agriculture in Uganda, Volume 1 General Information*, Kampala, Uganda: Fountain Publishers/CTA/NARO.

Müller, B., Höhne, N. and Ellerman, C. (2009) 'Differentiating (historic) responsibilities for climate change', *Climate Policy*, 9(6): 593–611.

Müller, M. (2007) 'What's in a word? Problematizing translation between languages', *Area*, 39(2): 206–13.

Myers, N. and Kent, J. (2003) 'New consumers: The influence of affluence on the environment', *Proceedings of the National Academy of Sciences*, 100(8): 4963–8.

Nair, C. (2011) *Consumptionomics: Asia's role in reshaping capitalism and saving the planet*, Singapore: John Wiley and Sons Asia.

ND-GAIN (Notre Dame Global Adaptation Initiative) (2019) ND-GAIN Country Index, Notre Dame, IN: University of Notre Dame, https://gain.nd.edu/our-work/country-index/

Newman, S. and Hatton-Yeo, A. (2008) 'Intergenerational learning and the contributions of older people', *Ageing Horizons*, 8: 31–9.

Nordström, P. (2016) 'The creative landscape of theatre-research cooperation: A case from Turku, Finland', *Geografiska Annaler: Series B, Human Geography*, 98(1): 1–17.

Norman, M. and Ueda, P. (2017) 'Biafran famine', in V.R. Preedy and V.B. Patel (eds) *Handbook of famine, starvation and nutrient deprivation*, Cham: Springer, pp 1–15.

Nuccitelli, D. (2016) 'The inter-generational theft of Brexit and climate change', *The Guardian*, 27 June, https://www.theguardian.com/environment/climate-consensus-97-per-cent/2016/jun/27/the-inter-generational-theft-of-brexit-and-climate-change

Oberheitmann, A. and Sternfield, E. (2011) 'Global governance, responsibility and a new climate regime', in P.G. Harris (ed) *China's responsibility for climate change: Ethics, fairness and environmental policy*, Bristol: Policy Press, pp 195–222.

Ochieng, R. (2003) 'Supporting women and girls' sexual and reproductive health and rights: The Ugandan experience', *Development*, 46(2): 38–44.

Olivier, J.G.J., Schure, K.M. and Peters, J.A.H.W. (2017) *Trends in global CO2 and total greenhouse gas emissions: Summary of the 2017 report*, Publication no. 2983, The Hague: PBL Netherlands Environmental Assessment Agency.

Osbahr, S., P. Dorward, R. Stern and S. Cooper (2011) 'Supporting agricultural innovation in Uganda to respond to climate risk: Linking climate change and variability with farmer perceptions', *Experimental Agriculture*, 47(2): 293–316.

Osborne, G. (2016) *Budget 2016: George Osborne's budget speech*, 16 March, www.gov.uk/government/speeches/budget-2016-george-osbornes-speech

Oxfam (2015) 'Extreme carbon inequality: Why the Paris deal must put the poorest, lowest emitting and most vulnerable people first', Oxfam Media Briefing, 2 December, www.oxfam.org/sites/www.oxfam.org/files/file_attachments/mb-extreme-carbon-inequality-021215-en.pdf

Page, E. (2006) *Climate change, justice and future generations*, Cheltenham: Edward Elgar Publishing.

Page, E. (2008) 'Distributing the burdens of climate change', *Environmental Politics*, 17(4): 556–75.

Parfit, D. (1984) *Reasons and persons*, Oxford: Oxford University Press.

Pearce, F. (2008) *Confessions of an eco-sinner: Travels to find where my stuff comes from*, London: Eden Project Books.

Peattie, K. and Collins, A. (2009) 'Perspectives on sustainable consumption', *International Journal of Consumer Studies*, 33(2): 107–12.

Pelling, M. (2011) *Adaptation to climate change: From resilience to transformation*, London: Routledge.

Pelling, M. and Wisner, B. (2009) *Disaster risk reduction: Cases from urban Africa*, London: Earthscan.

Pepper, M., Jackson, T. and Uzzell, D. (2009) 'An examination of the values that motivate socially conscious and frugal consumer behaviours', *International Journal of Consumer Studies*, 33(2): 126–36.

Persson, I. and Savulescu, J. (2012) *Unfit for the future: The need for moral enhancement*, Oxford: Oxford University Press.

Phillips, L.G. and Bunda, T. (2018) *Research through, with, and as storying*, London: Routledge.

Phoenix, A., Boddy, J., Walker, C. and Vennam, U. (2017) *Environment in the lives of children and families*, Bristol: Policy Press Shorts.

Pickett, K. and Wilkinson, R.G. (2009) *The Spirit Level: Why Equality is Better for Everyone*, London: Bloomsbury.

Pieterse, E. (2006) 'Building with ruins and dreams: Some thoughts on integrated urban development in South Africa through crisis', *Urban Studies*, 43(2): 285–304.

Podoshen, J.S., Li, L. and Zhang, J. (2011) 'Materialism and conspicuous consumption in China: A cross-cultural examination', *International Journal of Consumer Studies*, 35(1): 17–25.

Popke, J. (2007) 'Geography and ethics: Spaces of cosmopolitan responsibility', *Progress in Human Geography*, 31(4): 509–18.

Porter, G.K. Hampshire, A. Abane, E. Robson, A. Munthali, M. Mashiri and A. Tanle (2010) 'Moving young lives: Mobility, immobility and inter-generational tensions in urban Africa', *Geoforum*, 41(5): 796–804.

Pöyliö, H. and Kallio, J. (2017) 'The impact of education and family policies on intergenerational transmission of socioeconomic status in Europe', in J. Erola and E. Kilpi-Jakonen (eds) *Social inequality across the generations*, Cheltenham: Edward Elgar Publishing, pp 204–24.

Pun, N. and Lu, H. (2010) 'Unfinished proletarianization: Self, anger, and class action among the second generation of peasant-workers in present-day China', *Modern China*, 36(5): 493–519.

Punch, S. (2002) 'Youth transitions and interdependent adult-child relations in rural Bolivia', *Journal of Rural Studies*, 18(2): 123–33.

Raupach, M.R., Marland, G., Ciais, P., Le Quéré, C., Canadell, J.G., Klepper, G. and Field, C.B. (2007) 'Global and regional drivers of accelerating CO2 emissions', *Proceedings of the Natural Academy of Sciences*, 104(24): 10288–93.

Richardson, M. (2015) 'Theatre as safe space? Performing intergenerational narratives with men of Irish descent', *Social and Cultural Geography*, 16(6): 615–33.

Rigg, J. and Ritchie, M. (2002) 'Production, consumption and imagination in rural Thailand', *Journal of Rural Studies*, 18(4): 359–71.

Robins, N. (1999) 'Making sustainability bite: Transforming global consumption patterns', *The Journal of Sustainable Product Design*.

Rowlingson, K., Joseph, R. and Overton, L. (2017) *Inter-generational financial giving and inequality: Give and take in 21st century families*, Basingstoke: Palgrave Macmillan.

Rudiak-Gould, P. (2012) 'Promiscuous corroboration and climate change translation: A case study from the Marshall Islands', *Global Environmental Change*, 22(1): 46–54.

Rudiak-Gould, P. (2013) ' "We have seen it with our own eyes": Why we disagree about climate change visibility', *Weather, Climate and Society*, 5(2): 120–32.

Rudiak-Gould, P. (2014) 'Climate change and accusation: Global warming and local blame in a small island state', *Current Anthropology*, 55(4): 365–86.

Samson, J., Bereaux, D., McGill, B.J., and Humphries, M.M. (2011) 'Geographic disparities and moral hazards in the predicted impact of climate change on human populations', *Global Ecology and Biogeography*, 20(4): 532–44.

Satterthwaite, D. (2009) 'The implications of population growth and urbanization for climate change', *Environment and Urbanization*, 21(2): 545–67.

Schucher, G. (2017) 'The fear of failure: Youth employment problems in China', *International Labour Review*, 156(1): 73–98.

Shanahan, H. and Carlsson-Kanyama, A. (2005) 'Interdependence between consumption in the North and sustainable communities in the South', *International Journal of Consumer Studies*, 29(4): 298–307.

Shao, A.T. and Herbig, P. (1994) 'Marketing implications of China's 'Little emperors'', *Review of Business*, 16(1): 16–21.

Shek, D. (2006) 'Chinese family research: Puzzles, progress, paradigms, and policy implications', *Journal of Family Issues*, 27(3): 275–84.

Sheridan, M.J. (2012) 'Global warming and global war: Tanzanian farmers' discourse on climate and political disorder', *Journal of Eastern African Studies*, 6(2): 230–45.

Shove, E. (2003) *Comfort, cleanliness + convenience*, Oxford: Berg.

Shove, E. (2010) 'Beyond the ABC: Climate change policy and theories of social change', *Environment and Planning A*, 42(6): 1273–85.

Shove, E., Pantzar, M. and Watson, M. (2012) *The dynamics of social practice: Everyday life and how it changes*, London: Sage.

Skillington, T. (2018) *Climate change and intergenerational justice*, Abingdon: Taylor & Francis.

Slegers, M.F.W. (2008) '"If only it would rain": Farmers' perception of rainfall and drought in semi-arid central Tanzania', *Journal of Arid Environments*, 72(11): 2106–23.

Slingerland, E. (2003) *Confucius analects: With selections from traditional commentaries*, Indianapolis, IN: Hackett Publishing Company.

Smith, D.M. (2000) *Moral geographies: Ethics in a world of difference*, Edinburgh: Edinburgh University Press.

Smith, F. (1996) 'Problematising language: Limitations and possibilities in 'foreign language' research', *Area*, 28(2): 160–6.

Smith, P.B. and Fischer, R. (2008) 'Acquiescence, extreme response bias and culture: A multilevel analysis', in F.J.R. van de Vijver, D.A. van Hemert and Y.H. Poortinga (eds) *Multilevel analysis of individuals and cultures*, New York: Taylor & Francis Group/Lawrence Erlbaum Associates, pp 285–314.

Sommers, M. (2010) 'Urban youth in Africa', *Environment and Urbanization*, 22(2): 317–32.

Sommers, M. (2011) *Stuck: Rwandan youth and the struggle for adulthood*, Athens and London: University of Georgia Press.

Spaargaren, G. and Oosterveer, P. (2010) 'Citizen-consumers as agents of change in globalizing modernity: The case of sustainable consumption', *Sustainability*, 2(7): 1887–908.

Stamm, K.R., Clark, F. and Eblacas, P.R. (2000) 'Mass communication and public understanding of environmental problems: The case of global warming', *Public Understanding of Science*, 9(3): 219–37.

Steinig, S. and Butts, B. (2009) 'Generations going green: Intergenerational programs connecting young and old to improve our environment', *Generations*, 4(6): 64–9.

Stephen Parker, R., Hermans, C.M. and Schaefer, A.D. (2004) 'Fashion consciousness of Chinese, Japanese and American teenagers', *Journal of Fashion Marketing and Management*, 8(2): 176–86.

Sun, G., D'Alessandro, S. and Johnson, L. (2014) 'Traditional culture, political ideologies, materialism and luxury consumption in China', *International Journal of Consumer Studies*, 38(6): 578–85.

Tamale, S. (2003) 'Out of the closet: Unveiling sexuality discourses in Uganda', *Feminist Africa*, 2: 42–9.

Tempest, S. (2003) 'Intergenerational learning: A reciprocal knowledge development process that challenges the language of learning', *Management Learning*, 34(2): 181–200.

Thøgersen, S. (2006) 'Beyond official Chinese: Language codes and strategies', in M. Heimer and S. Thøgersen (eds) *Doing fieldwork in China*, Honolulu: University of Hawaii Press, pp 110–26.

Thompson, H. (2017) 'It's still the 2008 crash', *Political Quarterly*, 88(3): 391–4.

Triandis, H.C. (1995) *Individualism and collectivism*, Boulder, CO: Westview Press.

Tschakert, P. (2007) 'Views from the vulnerable: Understanding climatic and other stressors in the Sahel', *Global Environmental Change*, 17(3-4): 381–96.

Tuan, Y.-F. (1989) *Morality and imagination: Paradoxes of progress*, Madison: University of Wisconsin Press.

UN General Assembly (2015) *Transforming our world: The 2030 Agenda for Sustainable Development*, 21 October 2015, A/RES/70/1.

UN/DESA (UN Development Policy and Analysis Division) (2018) 'Least Developed Country Category: Uganda profile', www.un.org/development/desa/dpad/least-developed-country-category-uganda.html

UNDP (United Nations Development Programme) (2015) 'Uganda targets 22% emissions cut to achieve low-carbon growth', UNDP Press Centre, 16 November, www.ug.undp.org/content/uganda/en/home/presscenter/articles/2015/11/16/uganda-targets-22-emission-cuts-to-achieve-low-carbon-growth-by-2030.html

UNEP (United Nations Environment Programme) (2017) *Emissions gap report 2017: A UN Environment synthesis report*, Nairobi, Kenya: UNEP.

Valentine, G. (2003) 'Boundary crossings: Transitions from childhood to adulthood', *Children's Geographies*, 1(1): 37–52.

Vandenbergh, M.P. (2008) 'Climate change: The China problem', *Southern California Law Review*, 81(5): 905–50, https://ssrn.com/abstract=1126685

Vanderbeck, R. (2007) 'Intergenerational geographies: Age relations, segregation and re-engagements', *Geography Compass*, 1(2): 200–21.

Vanderbeck, R.M. and Worth, N. (2014) 'Introduction', in R.M. Vanderbeck and N. Worth (eds) *Intergenerational space*, Abingdon/New York: Routledge, pp 1–15.

Vedwan, N. (2006) 'Culture, climate and the environment: Local knowledge and perception of climate change among apple growers in northwestern India', *Journal of Ecological Anthropology*, 10(1): 4–18.

Walker, G. (2011) *Environmental justice: Concepts, evidence and politics*, London: Routledge.

Wallis, G. (1970) 'Chronopolitics: The impact of time perspectives on the dynamics of change', *Social Forces*, 49(1): 102–08.

Wang, B., Shen, Y. and Jin, Y. (2017) 'Measurement of public awareness of climate change in China: Based on a national survey with 4,025 samples', *Chinese Journal of Population Resources and Environment*, 15(4): 285–91.

Wang, C.L. (2009) 'Little Emperors: The future of China's consumer market', *Young Consumers*, 10(2), doi.org/10.1108/yc.2009.32110baa.001.

Ward, B. and Hicks, N. (2018) 'What is the polluter pays principle?', Grantham Research Institute at LSE, 11 May, www.lse.ac.uk/GranthamInstitute/faqs/what-is-the-polluter-pays-principle/

Warde, A. (2005) 'Consumption and theories of practice', *Journal of Consumer Culture*, *5*(2): 131–53.

WCED (World Commission on Environment and Development) (1987) *Our common future: Report of the World Commission on Environment and Development*, G.H. Brundtland (ed), Oxford: Oxford University Press.

Weber, C.L., Peters, G.P., Guan, D. and Hubacek, K. (2008) 'The contribution of Chinese exports to climate change, *Energy Policy*, 36(9): 3572–7.

Werner, O. and Campbell, D.T. (1970) 'Translating, working through interpreters, and the problem of decentering', in R. Naroll and R. Cohen (eds) *A handbook of cultural anthropology*, New York: American Museum of National History, pp 398–419.

Wexler, L. (2011) 'Intergenerational dialogue exchange and action: Introducing a community-based participatory approach to connect youth, adults and elders in an Alaskan native community', *International Journal of Qualitative Methods*, 10(3): 248–64.

White, J. (2017) 'Climate change and the generational timescape', *Sociological Review*, 65(4): 763–78.

Whitmarsh, L. (2009) 'What's in a name? Commonalities and differences in public understanding of 'climate change' and 'global warming'', *Public Understanding of Science*, 18(4): 401–20.

Willetts, D. (2010) *The pinch: How the Baby Boomers took their children's future – and why they should give it back*, London: Atlantic Books.

World Bank (2014) 'CO2 emissions (metric tons per capita)', https://data.worldbank.org/indicator/EN.ATM.CO2E.PC

Wu, S., Dai, E., Zheng, D. and Yang, Q. (2007) 'Case study on environmental ethics in sustainable development: Responsibilities for different communities', *Geographical Research*, 26(6): 1109–16. [吴绍洪, 戴尔阜, 郑度, 杨勤业 (2007) 区域可持续发展中的环境伦理案例分析：不同社会群体责任, *地理研究* 26(6): 1109–16.]

Wu, X. and Treiman, D.J. (2007) 'Inequality and equality under Chinese socialism: The Hukou system and intergenerational occupational mobility', *American Journal of Sociology*, 113(2): 415–45.

Yan, Y. (2009) *The individualization of Chinese society*, Oxford: Berg.

Yan, Y. (2010) 'The Chinese path to individualization', *The British Journal of Sociology*, 61(3): 489–512.

Young, I. (2003) 'From guilt to solidarity: Sweatshops and political responsibility', *Dissent,* 50: 39–44.

Yu, L. (2014) *Consumption in China: How China's new consumer ideology is shaping the nation*, Cambridge: Polity Press.

Zhang, Y. and Ruan, P. (2005) 'The study on the measurement and solutions of intergeneration fair problems', *Scientific Management Research*, 23(4): 25–8. [张勇, 阮平南 (2005) "代际公平"问题的测定和对策研究. 科学管理研究 23(4): 25–8.]

Index

D

H

I

M

N

O

P

Q

R

S

T

U